W9-CHH-130

BIOASSAY

Second Edition
Second Printing

John J. Hubert

Department of Mathematics
and Statistics
University of Guelph

**KENDALL/HUNT
PUBLISHING COMPANY**
Dubuque, Iowa

B   403953   02

Dedicated to
Jody and Jason

Contents

Preface to the Second Edition

This second edition has incorporated the changes suggested by the encouraging reviews of the first edition. I have added over 300 new references in the bibliography which indicates the growing level of research in bioassay in the last five years. An index has been included; several new examples and sections have been added; the questions of chapter 7 have been replaced with a description of a multivariate approach to time dependent quantal assays; and the technical and typographical errors have been eliminated. I am sincerely grateful to all the reviewers who pointed out the shortcomings of the first edition.

Since the first edition, I have developed more computer software to the existing programs such that all analyses can be done on a microcomputer. A floppy disc now contains almost all the necessary programs. With the adoption of this monograph for course use, the instructor can obtain, free of charge, this software by writing to the author; otherwise, for a small charge, I could duplicate it for those researchers in bioassay.

J. J. H.
Mar. 20, 1984

This second printing of the second edition has an enlarged bibliography and an index to this new list. Also, the typographical errors have been eliminated.

J. J. H.
Apr. 15, 1986

''I think it consists of the capacity to learn. It is an attitude of mind which is never satisfied with its representation of the world; a mind which invites fact to suggest theory, however unwelcome, and requires theory to discover fact which can conform or discredit opinion. With this you should have no difficulty in solving your problems.''

G. E. P. Box, May 11, 1975.

J. J. H.
1979

1 Introduction

As you have already learned, statistics is both a science and an art. Because statistics utilizes mathematical principles in order to describe and compare populations via parameters which characterize certain attributes of the populations, it is considered a science. It is however recognized that particular assumptions which are seldom valid in the real world are the basis for these mathematical considerations. When we approach data by appealing to the experience of others, to our guesses about assumptions and to strict experimental rules, we are attempting to be objective. This can be considered an art.

In any case, once data have been obtained, applying statistics usually follows these steps: casting the experimental problem into a statistical problem, choosing the appropriate statistical method, applying the procedure, obtaining statistical answers, and finally, interpreting these answers back into the original experimental problem.

Biostatistics (or biological statistics, biometry) is the application of statistical methods to the solution of biological problems. Within biostatistics we have *bioassay,* a body of procedures in which the amount or strength of an agent or stimulus is determined by a response of a subject. The *subject* is usually an animal, a human tissue, or a bacterial culture; the *agent* is usually a drug; and the *response* is usually a change in a particular characteristic or even death of a subject. The estimation of the nature or potency of the agent by the response will be one of the primary objectives.

Webster's International Dictionary defines a biological assay as "the estimation of the strength of a drug, etc., by comparing its effect on biological material, as animals or animal tissue, with those of a standard product." If X_t units of the test drug perform like ρX_s units of the standard (i.e., the test acts like a dilution of the standard), then the relative potency, denoted by ρ (rho), of an agent to some standard agent is the ratio of the effective constituent in a unit amount of the test preparation to that of the standard. For example, if $\rho = 2$, this means one unit of the test preparation is equivalent to 2 units of the standard preparation (or ½ unit of test equals 1 unit of standard) with respect to some biological activity.

There are several types of assays. Schematically,

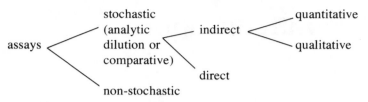

Non-stochastic assays assume ρ does not depend on the subject, i.e., on the species of the animal, the route of administration, laboratory differences, etc., but rather that ρ is some constant and is deterministic. Such assays are not stochastic and are not analyzed in this textbook. An assay is *stochastic* if ρ is influenced by factors other than the preparations, i.e., extraneous (e.g., environmental) factors which cannot be completely controlled.

Analytic dilution assays are such that the test and standard preparations behave as though they were identical, except for the concentration of the substance being assayed. *Comparative* assays do not have the condition of analytic assays; for comparative assays it is known that the preparations are not the same. For example, the concentration of one protein is estimated by using a different protein as a standard, or the potency of a new drug relative to a different standard drug is wanted.

In the *direct* assay we measure the amount of stimulus (e.g., concentration of a drug) holding the response fixed. In particular if a drug is used to kill an animal, we can vary the dose till the heart beat ceases; in this way we can directly observe the response and we can measure directly the amount (or concentration) of dose used. In direct assays the dose is the variable of interest, not the response.

In *indirect* assays we observe a variable related to an event; for example we observe the tension in a tissue, the hormone level or the blood sugar content, and this is related to the event under study. This is the most common type of assay and the most interesting statistically. In *qualitative* assays every experimental unit obtains the same dose and we only observe proportions or degrees of response and often measurements are not even possible; in fact responses may only be death. Here the number of responses is the variable of interest. An example of a quantal (binary, or all-or-nothing) type of assay is: after injection with different doses of insulin, the percentage of mice showing convulsions at each dose level is recorded. Quite simply, quantal assays involve the investigation of the relation between dosage and response percentages.

In *quantitative* indirect assays measurements are possible and here too it is response rather than level of dosage which is the variable of interest.

In addition to the books by Finney (1964), (1971) and Bliss (1952), other books containing bioassay are Armitage (1971), Dawn et al. (1962), Colquhoun (1971), Federer (1973) and Parker (1973). Also helpful are the papers by Brown (1966), Das et al. (1966), Finney (1947), Irwin (1937), (1950), and Jerne and Wood (1949). Early excellent reviews of the physical and biological experimental procedures can be found in Busvine (1957), Sun (1957), Nagasawa (1959) and Hoskins and Craig (1962). References for statistical prerequisites are: Goldstein (1964), Wadley (1967), Raktoe and Hubert (1979), Sokal and Rohlf (1969), Malik and Mullen (1973), Campbell (1974), Daniel (1974), and Zar (1974).

2 Direct Assays

Direct assays primarily involve a quantity known as the *individual effective dose* (IED) which is defined as the amount (or concentration, dose) of a drug required to produce a particular response (e.g., cardiac arrest) in a particular subject (e.g., cat, fish). A typical experiment on a group of subjects begins by dividing randomly this group into two groups and then the IED of a standard preparation is measured on each subject of one group; the IED of the test or unknown preparation is measured on each subject of the other group. In this way such assays can be used to estimate the relative potency, i.e., the ratio of concentrations of the test to standard to produce the same biological effect. In particular, a standard preparation of a particular drug, and a preparation of the same drug of unknown concentration can be compared and used to estimate the unknown concentration. In direct assays the variable of interest is the dose required to produce a response whereas in indirect assays the variable of interest is the response and not the dose levels.

In this chapter we explain two estimators for the relative potency for direct assays: the ratio of means estimator and the ratio of standard deviations estimator. There are two other methods, the Shorack (1966) graphical method and the Rao-Littel (1976) nonparametric method for finding confidence limits, which are elegant refinements but will not be discussed here.

2.1 Example 2.1

A direct assay of the lethal drug ''strophanthus'' involved the infusion of this agent into cats. For a standard preparation and an unknown (test) preparation of this drug, the amount of this drug (actually the amount in c.c. per kg. of body weight of the cat) needed to produce cardiac arrest was recorded. A portion of the results of this experiment is contained in Table 2.1. (This is part of a more involved experiment first reported in Burn et al. (1950) and later analyzed in Finney (1964).)

4

Table 2.1
THE STROPHANTHUS DATA

	Standard X_s	Test X_t
	2.42	1.55
	1.85	1.58
	2.00	1.71
	2.27	1.44
	1.70	1.24
	1.47	1.89
	2.20	2.34
Total	13.91	11.75
Mean	1.9871	1.6786
ΣX^2	28.3227	20.4819
Variance	0.11359	0.12645

The objective of this experiment was to estimate the relative potency. We shall now find the point and interval estimates.

2.2   *The ratio of means estimator*

The *ratio of means point estimate* of the relative potency, ρ, is given by

$$\hat{\rho} = R = \frac{1/\overline{X}_t}{1/\overline{X}_s} = \frac{\overline{X}_s}{\overline{X}_t} = \frac{1.987}{1.679} = 1.18. \qquad (2.1)$$

Thus one unit of the test preparation is estimated to be equivalent to 1.18 units of the standard. We can interpret $\overline{X}_t$ to be an estimate of the average amount of the test preparation required to kill a cat. Recall, ρ is also the number of units of effective constituent contained in (or equivalent to) a unit dose of the test preparation; thus if it was known that the unknown (test) preparation was, for example, twice as potent as the standard then only half as much of the test will be needed to produce the same effect as the standard, so that $\rho = 2$. By the way, it can be shown that the ratio of means estimator R is a consistent estimator of ρ. Notice that we are defining relative potency by the relation $\rho = \mu_s/\mu_t$. We can also define relative potency by $X_s = \rho\, X_t$; that is, if X_t units of the test preparation will kill an animal then $X_s = \rho\, X_t$ units of the standard preparation will kill the same animal. Furthermore, if we assume that the distribution of the lethal doses of the standard preparation over animals to have mean $E[X_s] = \mu$ and variance $V[X_s] = \sigma^2$, and if the same animals were killed with the test preparation, then the distribution of the lethal doses X_t would possess a distribution with $E[X_t] = \mu/\rho$ and $V[X_t] = \sigma^2/\rho^2$.

To obtain the *interval estimate* for ρ we would require the sampling distribution of the statistic R. However, since the mathematical statistics involved in the derivation of the confidence limits for ρ is rather involved, we will here only state the result which we will need and show how and when to use it. The following result is a consequence of the so-called Fieller Theorem (see Fieller (1940), (1944), (1947) and also Geary (1930), Creasy (1956), and Elston (1969)); for completeness' sake the statement and proof are given in Appendix 1. The proof of the following theorem is given in Appendix 2.

Theorem 2.1   Limits for ρ: normal case

If the populations are normally distributed with equal variability and if the samples are random and independent, then the confidence limits for the true relative potency is given by

$$
\frac{R \pm t \, \dfrac{S}{\overline{X}_t} \sqrt{\dfrac{1}{n_s}(1 - g) + \dfrac{R^2}{n_t}}}{1 - g},
\tag{2.2}
$$

where R is the sample relative potency $= \overline{X}_s / \overline{X}_t$, S^2 is the pooled variance of the 2 samples, i.e.,

$$
S^2 = \frac{(n_s - 1) S_s^2 + (n_t - 1) S_t^2}{n_s + n_t - 2},
\tag{2.3}
$$

$$
g = \frac{t^2 \, S^2}{n_t \, \overline{X}_t^2},
\tag{2.4}
$$

and t is the appropriate Student's t-value based on degrees of freedom (d.f.) $n_s + n_t - 2$ (e.g., $t_{.975}(12) = 2.179$).

The following result provides an approximation to Theorem 2.1.

Corollary 2.1   Special Case

If the sample sizes are equal, i.e., $n_t = n_s = n$, say, and if the quantity g is relatively small, then an approximate form of the confidence limits for ρ is

$$
R \pm t \frac{S}{\overline{X}_t} \sqrt{\frac{1 + R^2}{n}}
\tag{2.5}
$$

where

$$
S^2 = \frac{1}{2}(S_s^2 + S_t^2).
\tag{2.6}
$$

Example 2.2   Continuation of 2.1

For the data of Table 2.1 we have

$n_t = n_s = 7$

$\overline{X}_t = 1.6786$

$R = 1.1838$

$t = t_{.975}(12) = 2.179$

$S = \sqrt{\dfrac{6(.11359) + 6(.12645)}{12}} = 0.34644$

$g = \dfrac{1}{7}\left(\dfrac{(2.179)(0.34644)}{1.6786}\right)^2 = 0.02899.$

Substitution into (2.2) yields the 95% confidence limits, 1.219 ± 0.270. Thus the 95% confidence interval is (0.95, 1.49), provided the assumptions of Theorem 2.1 hold.

It is interesting to note that the approximate limits given by (2.5) give 1.184 ± 0.263, so that the 95% confidence interval becomes (0.92, 1.45) which is to the left of the more accurate result.

The above calculations hold provided the three basic statistical assumptions are valid. (The normality of the population of X values could be tested by the χ^2 goodness of fit test if the samples were large, homoscedasticity by the F test, and the independence of the samples would be supported by proper randomization.)

There is another formula for the interval estimator for ρ which has been suggested by Brown (1964): when the sample sizes are large or when the *coefficient of variation* is small then the approximate 95% confidence interval for ρ is

$$R \pm 1.96\sqrt{\frac{1}{\overline{X}_t^2}\left(\frac{S_s^2}{n_s} + R^2\frac{S_t^2}{n_t}\right)}. \tag{2.7}$$

The expression under the square root sign arises from the following argument. For any random variable X, say, the coefficient of variation, $CV[X]$, is defined as $CV[X] = V[X]^{1/2}/E[X]$. It is also known that for arbitrary random variables X and Y we have $(CV[X/Y])^2 \simeq (CV[X])^2 + (CV[Y])^2$. Thus for our variables we have

$$\frac{V[R]}{(E[\overline{X}_s/\overline{X}_t])^2} = \frac{V[\overline{X}_s]}{(E[\overline{X}_s])^2} + \frac{V[\overline{X}_t]}{(E[\overline{X}_t])^2}$$

$$\frac{V[R]}{\rho^2} = \frac{V[\overline{X}_s]}{\rho^2\mu_t^2} + \frac{V[\overline{X}_t]}{\mu_t^2}$$

$$V[R] = \frac{1}{\mu_t^2}\left\{\frac{\sigma_s^2}{n_s} + \rho^2\frac{\sigma_t^2}{n_t}\right\}.$$

Therefore an estimate of this variance can be approximated by

$$\hat{V}[R] = \frac{1}{\overline{X}_t^2}\left\{ \frac{S_s^2}{n_s} + R^2\, \frac{S_t^2}{n_t} \right\}. \tag{2.8}$$

There is another result which may be helpful if the assumptions of Theorem 2.1 are not valid. The proof of the following result is given in Appendix 3.

Theorem 2.2   Limits for ρ: log-normal case

If the two populations are log-normal, (that is, if $Z_t = \log X_t$ and $Z_s = \log X_s$ are normally distributed) with equal variability and if the two samples are independent, then the confidence limits for the relative potency, ρ, are given by

$$\text{antilog}\left(R^* \pm t\, S\, \sqrt{\frac{1}{n_t} + \frac{1}{n_s}} \right), \tag{2.9}$$

where $R^* = \overline{Z}_s - \overline{Z}_t$, and t and S are defined as in Theorem 2.1, using the data of the Z variables.

Example 2.3

Under logarithmic transformation suggested by Theorem 2.2, the data of Table 2.1 is transformed into Table 2.2. The pooled variance is $S^2 = 0.006818$, and $R^* = 0.0755$. The 95% confidence limits are given by $10^{.0755 \pm .0962}$ which yields

$$(10^{-0.0207}, 10^{+0.1717}) = (0.953, 1.485).$$

Hence the 95% confidence interval for ρ is (0.95, 1.49).

Table 2.2

	Z_s	Z_t
	.3838	.1903
	.2672	.1987
	.3010	.2330
	.3560	.1584
	.2304	.0934
	.1673	.2765
	.3424	.3692
Sum	2.0481	1.5195
Mean	0.2926	0.2171
ΣZ^2	0.6344	0.3765
Variance	0.00585	0.00779

2.3   The ratio of standard deviations estimator

Another approach to estimate the relative potency is referred to as the *ratio of standard deviations* approach. This approach is based on the defining relation $X_s = \rho\, X_t$, whereas the ratio of means estimator is based on the defining relation $\rho = \mu_s/\mu_t$.

If we assume the lethal dose distribution of the standard preparation over subjects has $E[X_s] = \mu_s$ and $V[X_s] = \sigma_s^2$ then the lethal dose distribution of the test preparation over the same subjects would have $E[X_t] = \mu_s/\rho$ and $V[X_t] = \sigma_s^2/\rho^2$, since $X_s = \rho\, X_t$. Thus $\rho = (V[X_s]/V[X_t])^{1/2}$ and an estimator, call it R', would be given by

$$\hat{\rho} = R' = S_s/S_t. \tag{2.10}$$

It can be shown that the estimate of the asymptotic variance is approximately

$$\hat{V}(R') = \frac{1}{S_t^2}\left\{ \frac{S_s^2}{2n_s} + R'^2\,\frac{S_t^2}{2n_t} \right\} = \frac{R'^2}{2}\left(\frac{1}{n_s} + \frac{1}{n_t} \right).$$

If we assume that X_s (and consequently X_t) has a normal distribution it can be shown that the exact 95% confidence limits for the relative potency is

$$(R'\sqrt{F_{.025}(n_t - 1, n_s - 1)},\ R'\sqrt{F_{.975}(n_t - 1, n_s - 1)}). \tag{2.11}$$

(Note that since $(n_s - 1)S_s^2/\sigma_s^2$ is $\chi^2_{(n_s-1)}$ and $(n_t - 1)S_t^2/(\sigma_s^2/\rho^2)$ is $\chi^2_{(n_t-1)}$ then $\rho^2 S_t^2/S_s^2$ is $F(n_t - 1, n_s - 1)$.)

For the data of example 2.1 the formula (2.11) yields the limits $(0.948\sqrt{.172},\ 0.948\sqrt{5.82}) = (0.39, 2.29)$.

An approximate 95% confidence interval for ρ based on R' is

$$R'\left(1 \pm \frac{1.96}{\sqrt{2}}\sqrt{\frac{n_s + n_t}{n_s n_t}} \right). \tag{2.12}$$

If we had defined ρ by the relation $X_s = \rho\, X_t$ then for example 2.1 we have from (2.10) R' = 0.948 and from (2.12) we obtain the approximate limits $0.948(1 \pm .741)$ or $(0.25, 1.65)$. The limits are much wider than before and this is characteristic of this type of estimator.

2.4   Conclusions

We have seen that for direct assays there are two different point estimators for the relative potency. The ratio of means estimator appears to be the most appropriate for several reasons: it is a natural interpretation of the biologist's view of relative potency, it is the most widely used estimator, in practice its interval estimator has smaller width and besides the exact confidence limits there are two convenient approximations (see (2.5) and (2.7)) and another form based on the log-normal distribution.

2.5 *Exercises*

2.1 For the direct assay on the tolerance of cats (see Example 2.1, 2.2):

   (a) From the point estimate of the relative potency, how many units of the test preparation is equivalent to 1 unit of the standard preparation with respect to the biological activity under investigation?

   (b) In order to find the interval estimate certain assumptions are necessary. (See Theorem 2.1). List these assumptions.

   (c) One of these assumptions was homogeneity of variance. Test this assumption.

   (d) Verify that the confidence interval is (0.95, 1.49) by substituting the values into (2.2) and (2.9).

   (e) Can one interpret the confidence interval found in (d) as "1 unit of the test preparation is not less potent than 0.95 units of the standard, and not more potent than 1.49 units of the standard preparation at the 95% confidence level"?

2.2 The data below were reported by Finney (1964):

	Preparations	
	Standard (A)	Test (B)
	2.4	5.2
	1.9	8.0
	2.0	4.8
	2.3	6.5
	1.7	7.0
		8.1
		6.0
Sum	10.3	45.6
Raw SS	21.55	306.94
Variance	0.083	1.648

This data shows the doses or tolerances of two groups of mice for two preparations of insulin, labelled A and B. Notice the different sample sizes. (The two batches of mice were as homogeneous as possible.) The tolerances were recorded as quantities (c.c.) per 100 grams body weight of mouse. (The tolerances are thus assumed to vary in proportion to body weight, or at least to show an approximately proportional variation rather than independence of body weight.) Provided that the mice have been assigned at random to the preparations estimate the relative potency and interpret this value.

Test for homoscedasticity so that Theorem 2.1 can be applied. If the variances are not significantly different apply this result. If the variances are significantly different apply Theorem 2.2. Interpret your results. Also apply (2.7) for comparison.

2.3 If $\overline{X}_t$ is greater than $\overline{X}_s$ do you expect R > 1? Why?

2.4 Is R* = log R? Why?

2.5 Does R require any distributional assumptions?

2.6 For the data in exercise 2.2 find the ratio of standard deviations estimator. Also find the 95% confidence limits by using the approximate formula (2.12) and by using the exact formula (2.11) which requires the normality assumption.

2.7 Suppose the null hypothesis is H: $\rho = 1$. For Example 2.1 and exercise 2.2, is there evidence to reject H at the 5% level of significance? What does the answer imply?

2.8 The following data, abridged from Asano (1961), on three short acting narcotics, a derivative A of *thiopentalsoda* and two derivatives of *methylhexabitalsoda* B and C. The preparation C was regarded as the standard preparation and A and B were compared with C. The subjects were "homogeneous puppies selected with care from those of the same ages and the same body weight, approximately". The assay was performed as follows: "a certain dosage of C was first infused as a preliminary operation and secondly a dosage of A or B is infused into a subject until a specified depth III of Girndt's measurement was observed". In the data the amount of dosage infused until depth III of anesthesia was observed.

C	A	B
21.0	18.0	35.5
26.0	13.0	39.0
19.0	13.5	38.5
16.0	11.5	37.0
22.0	15.0	34.0
104.0	71.0	184.0

(a) Estimate the relative potency of B to C and A to C using the estimators (2.1) and (2.10).

(b) Find the 95% confidence interval for ρ using

(i) (2.2) if possible, and if not, use (2.9), if possible.

(ii) (2.7).

(iii) (2.11) and (2.12).

(c) Comment on all of your answers.

(d) Is there any difference between A and B relative to C?

3 The Indirect Quantitative Assays

In an *indirect* assay specified doses are given each to several subjects and the nature of their responses is recorded. The record for each test may state merely that a characteristic response such as death is or is not produced: this is a *quantal* or *all-or-nothing* response. Alternatively, the magnitude of some property of the subject, its weight, the weight of a particular organ, or its time of survival, may be measured: this is a *quantitative* response.

Many bioassays are very complicated and long; the common basic components of a statistical analysis of such experiments are:

a. Statement of problem and objectives (e.g. How potent is a test preparation relative to a standard?)
b. The design of the experiment (e.g. RCBD, 6 point symmetric layout)
c. Details of the controlled experiment and tabulation of data
d. Preliminary analysis (e.g. graphs)
e. Testing and accounting for differences (e.g. ANOVA)
f. Estimation of toxicity and/or relative potency
g. Confidence limits on parameters
h. Analysis of combination of independent assays
i. Interpretation, conclusions and recommendations.

We will illustrate the above components through a simple comparative assay. Other types of assays and indeed more complex assays will include more than these basic components.

In this section we analyze in detail a comparative indirect quantitative assay. Although the data is fictitious it is typical of this type of assay.

3.1 *Example 3.1 An indirect quantitative assay.*

When a pharmaceutical company wants to place a drug on the market, government authorization (involving inspection and control) is required. One aspect of the problem is to measure the drug's potency relative to a known (or standard) preparation.

The data of Table 3.1 is the result of a controlled experiment involving bacterial cultures utilizing a randomized complete block design. Suppose that the basic unit recorded was the amount of decrease in growth and that 8 dishes each containing 6 identical cultures were used. Notice that for the test and standard preparations of the agent under investigation the number of dose levels and the value of the dose levels are the same. (See exercise 3.3 for the results of three more replications of this experiment.) Table 3.1 also contains certain calculations which will be needed in subsequent sections.

When we expect an increase in response with an increase of dosage (e.g. when slopes are positive), the drug with the greatest response on the average should be more potent; that is, when $\overline{Y}_t > \overline{Y}_s$ we expect $R > 1$. This is a natural result for indirect assays. One should contrast this with the intuitive result that we expect $R > 1$ when $\overline{X}_t < \overline{X}_s$ in the case of direct assays wherein we fix response and measure the dosages.

The next stage of the analysis is to discover the relationship between the magnitude of response (Y) and level of dose (d). The reason we want to discover this relationship is because such a relationship will allow us to go from the measured response variable, Y, to the fixed dosage variable; and it is the dosage variable in terms of which the relative potency is defined. One could plot Y versus d, Y versus $X = \log d$ (to any base), Y versus $1/d$ and so on. However, history gives us the following result: plots of Y versus $X = \log_{10} d$ suggest straight line patterns which are parallel. As will be seen later the benefits of a parallel line plot are considerable.

For the data of our example the plot of Y versus d on semi-logarithmic paper is given in Figure 3.1. There are several interesting characteristics: (1) there are 6 points, (2) the values plotted are average of Y (mean response) against the corresponding value of X (on log axis this is the log dosage), (3) a suggested straight line pattern for both test and standard preparations, (4) the obvious parallelism of the two regression lines, (5) the clear suggestion that for any given response (a Y value) the inverse mapping to the X axis yields a difference in X (or log d) values and this value is the same for any fixed Y value.

There is a statistical argument which may support this parallelism. If the response random variable for the standard preparation, Y_s, has a linear regression with $X_s = \log d_s$, then we can let the conditional expectation of Y_s given X_s be

$$E[Y_s | X_s] = a + bX_s,$$

where we need only assume that the distribution of Y_s at each X_s has variance σ^2 and that all measurements Y_s are independently distributed. (To obtain confidence limits and to make tests of significance it will be necessary of course to make some distributional assumptions like normality.) If we assume a dose of d_t units of the test preparation is equivalent to a dose of $d_s = \rho \, d_t$ units of the standard for all dose levels, then necessarily $E[Y_t | \log d_t] = E[Y_s | \log (\rho \, d_t)]$; and since $X_s = \log d_s = \log (\rho \, d_t) = \log \rho + \log d_t = \log \rho + X_t$ then $E[Y_t | X_t] = a + b \log \rho + b \, X_t$ or $E[Y_t | X_t] = a' + b \, X_t$. Comparing the two regression models we see that the intercepts are different, a and $a' = a + b \log \rho$, but the slopes are the same.

Table 3.1

	STANDARD			TEST			
dose (d) (mg/cc)	.25	.50	1.0	.25	.50	1.0	
$\log_{10}$ dose (X)	$-.60206$	$-.30103$	0	$-.60206$	$-.30103$	0	
							Total
	4.9	8.2	11.0	6.0	9.4	12.8	52.1
	4.8	8.1	11.5	6.8	8.8	13.6	53.6
	4.9	8.1	11.4	6.2	9.4	13.4	53.4
response (Y)	4.8	8.2	11.8	6.6	9.6	13.8	54.8
(mm)	5.3	7.6	11.8	6.4	9.8	12.8	53.7
	5.1	8.3	11.4	6.0	9.2	14.0	54.0
	4.9	8.2	11.7	6.9	10.8	13.2	55.7
	4.7	8.1	11.4	6.3	10.6	12.8	53.9
sum	39.2	64.8	92.0	51.2	77.6	106.4	431.2
mean	4.9	8.1	11.5	6.4	9.7	13.3	9.8
ΣY^2	192.38	525.2	1058.5	328.5	756.0	1416.72	—

	STANDARD	TEST
$n_s = n_t = n$	24	24
$\overline{X}$	-0.30103	-0.30103
ΣX^2	3.6248	3.6248
$\Sigma(X - \overline{X})^2$	1.4499	1.4499
ΣY^2	1776.08	2501.22
$\overline{Y}$	8.17	9.80
$\Sigma(Y - \overline{Y})^2$	175.413	196.26
ΣXY	-43.1075	-54.1854
$\Sigma(X - \overline{X})(Y - \overline{Y})$	15.894	16.6169
$\hat{a}$	11.467	13.250
$\hat{b}$	10.962	11.461

It is important to remark that non-parallel line assays have been observed. Fortunately techniques have been developed to handle even these cases: for *parabolic* assays see Bliss (1957), Cornfield (1964), Elston (1965), Cox (1972) or Williams (1973); for *slope-ratio* assays see section 3.13; and for *asymmetric* assays see section 3.14.

The next stage of our analysis is to test if the lines are indeed parallel. Moreover, it will be necessary to statistically show that the lines are straight, that the lines do not have zero slope, and that the lines are not close together. We do this in sections 3.2 and 3.3 and we estimate the relative potency in section 3.4.

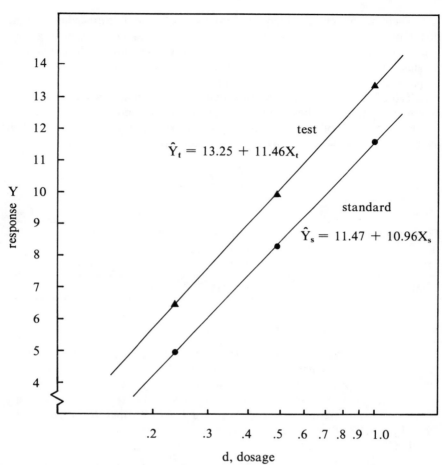

Figure 3.1. Quantitative parallel line assay responses against log doses.

3.2 *The ANOVA for treatment differences*

We are now at the testing stage. The data can be cast into the following design characteristics: 6 treatments, 8 blocks, 48 responses; the 8 blocks (replicates) of each of the six doses were given in sequence (in time) all to the same type of culture; the order in which the doses were arranged in each block was decided strictly at random using random number tables. It is apparent that to detect any treatment difference we should apply the ANOVA procedure for a randomized complete block design on the responses.

There are of course certain assumptions involved in this procedure: (1) the distribution of errors in the observation is normal, (2) equal scatter in all groups (i.e., homoscedasticity), (3) the size of response was not affected by previous or neighbouring responses (i.e., uncorrelated responses), (4) additivity. One should be prepared to test all of these assumptions.

Table 3.2 shows that there is a significant difference in the treatments. Geometrically, this implies that we can exclude the possibility that the two lines are on top of each other and parallel to the X axis. That is, we reject the null hypothesis $H : \mu_1 = \ldots = \mu_6$ using the usual ANOVA procedure.

Table 3.2
ANOVA TABLE

Source	d.f.	SS	MS	F_s	$F_{.95}$
Treatments	5	396.8667	79.3733	501.5	2.5
Blocks	7	1.2800	—	—	—
Error	35	5.5400	0.1583	—	—
Total	47	403.6867			

3.3 *The orthogonal linear contrast concept*

To explain the various components representing the different sources of variability in the treatment differences, we need the concept of an *orthogonal linear contrast*. We will explain this concept by using our example.

Let T_j denote the totals of the r responses to the jth dose level where $j = 1, 2, \ldots, J$. Here r = 8 and there are J = 6 distinct doses; and $T_1 = 39.2$, $T_2 = 64.8$, $T_3 = 92.0$, $T_4 = 51.2$, $T_5 = 77.6$ and $T_6 = 106.4$.

Let

$$L_i = \sum_{j=1}^{J} \alpha_{ij} T_j, \, i = 1, 2, \ldots, 5. \tag{3.1}$$

These five quantities are called the sample linear contrasts in the sense that each one is a linear combination of the 6 treatment totals (doses) and each one is an estimator of a corresponding contrast ϕ_i, say. The coefficients will have to be chosen so that (a) for each contrast

$$\sum_{j=1}^{J} \alpha_{ij} = 0 \tag{3.2}$$

and (b)

$$\sum_{j=1}^{J} \alpha_{ij} \cdot \alpha_{i'j} = 0, \, i \neq i', \, i, \, i' = 1, 2, \ldots, 5; \tag{3.3}$$

i.e., the sum of the coefficients of each contrast L_i is zero and the sum of the products of the coefficients of the corresponding totals T_j is zero for all possible pairs of contrasts L_i, the orthogonality property. The choice of the values of these coefficients when $J = 6$ is defined by Table 3.3 where the two properties (a) and (b) can readily be checked to be true.

Table 3.3
VALUES OF COEFFICIENTS (α_{ij}) FOR
CONTRASTS IN 6 POINT ASSAY

	T_1	T_2	T_3	T_4	T_5	T_6	$\Sigma\alpha$	$\Sigma\alpha^2$
L_1	-1	0	1	-1	0	1	0	4
L_2	1	0	-1	-1	0	1	0	4
L_3	-1	-1	-1	1	1	1	0	6
L_4	1	-2	1	1	-2	1	0	12
L_5	-1	2	-1	1	-2	1	0	12

The coefficients were not chosen by some mad scientist; they were in fact chosen for very special reasons. First, note that T_1 is the total corresponding to the lowest dose level and T_3 corresponds to the highest dose level for the standard preparation. If the slope of the line for the standard was zero then $T_3 - T_1$ would equal zero. Similarly, if the slope for the test preparation was zero then $T_6 - T_4$ would be equal to zero. Then ϕ_1, their sum, would be zero if the slopes were zero. We call ϕ_1 the *regression contrast* and if the slopes are not significantly different from zero, then ϕ_1 will not be significantly different from zero. Moreover if ϕ_1 were equal to zero then we would have no variation in response as we vary the level of concentrations, a very poor situation indeed!

The quantity ϕ_2 is called the *deviation from parallelism contrast* (if the lines are parallel then $\phi_2 = 0$); ϕ_3 is the *between standard and test preparation contrast* (if the mean response of T equals mean response of S then $\phi_3 = 0$; i.e., no difference in preparations with respect to response); ϕ_4 is the *deviations from linearity contrast* (if the true lines are straight, $\phi_4 = 0$); ϕ_5 is the *difference in curvature contrast* (if deviations from linearity are the same for both the standard and the test preparation then $\phi_5 = 0$). (Diagrams can be drawn to illustrate these properties.)

To calculate the sample contrasts we use the values of the totals (T_j) and the appropriate coefficients, the α_{ij} given in Table 3.3. For example,

$$L_1 = -T_1 + 0T_2 + T_3 - T_4 + 0T_5 + T_6$$
$$= -39.2 + 92.0 - 51.2 + 106.4 = 108.0$$

and

$$L_5 = -(39.2) + 2(64.8) - (92.0) + (51.2) - 2(77.6) + (106.4)$$
$$= 0.80.$$

The sum of squares associated with ith contrast L_i can be shown to be

$$SS(L_i) = \frac{L_i^2}{r\Sigma\alpha^2} \qquad (3.4)$$

where r = total number of observations per total, and $\Sigma\alpha^2$ for each L_i is given in the last column of Table 3.3. For example, $SS(L_5) = (0.8)^2/8(12) = 0.01$. Moreover, it is known that

$$\sum_{i=1}^{5} SS(L_i) = SS(\text{among treatments}) = SS(T) \qquad (3.5)$$

which explains how the treatment sum of squares is partitioned among the five contrasts. Notice that $SS(T)$ has 5 degrees of freedom and each of the sum of squares of the five contrasts has one degree of freedom.

Table 3.4 shows the values of each of the five contrasts and the associated sum of squares for the example we have been considering.

Table 3.4
VALUES OF CONTRASTS

L_i	Value	$SS = L_i^2/r\Sigma\alpha^2$
L_1	108.0	364.50
L_2	2.4	0.18
L_3	39.2	32.01
L_4	4.0	0.17
L_5	0.8	0.01
Total	—	396.87

We are now ready to summarize our analysis. The complete ANOVA for our example is given in Table 3.5. (Notice that we have included an "interpretation" column; this is for instructional purposes only.) The critical values are $F_{.95}(1,35) = 4.13$ and $F_{.95}(5,35) = 2.49$.

Since a significant regression and a significant difference between the preparations is reflected by large sums of squares for L_1 and L_3 respectively and since a significant parallelism, linearity and departure from quadratic regression is reflected by small sums of squares for L_2, L_4 and L_5 respectively, then the appropriate null hypotheses which we are testing in Table 3.5 for L_1 to L_5 are

$H_{(1)}$: slopes are zero (i.e., no regression)

$H_{(2)}$: lines are parallel

$H_{(3)}$: there is no difference between the two preparations

$H_{(4)}$: lines are straight

$H_{(5)}$: both lines deviate from linearity in the same fashion.

Table 3.5
COMPLETE ANOVA

Source	d.f.	SS	MS	F_s	Interpretation
L_1	1	364.50	''	2303.00	slope is not zero
L_2	1	0.18	''	1.14	lines are parallel
L_3	1	32.01	''	202.25	difference between responses
L_4	1	0.17	''	1.05	lines are straight
L_5	1	0.01	''	0.04	deviations from linearity same for standard and test
treatments	5	396.867	79.3733	501.5	treatments differ
blocks	7	1.280	—		
error	35	5.540	0.1583		
Total	47	403.687			

Finney (1964), p. 142, refers to ϕ_4 as the *quadratic curvature contrast* and ϕ_5 as the *difference in quadratics contrast*. The five contrasts form the necessary set for this 6-point symmetrical assay. For a detailed discussion of 4-point and 8-point symmetrical assays one should also refer to Finney. For completeness' sake, we provide in Table 3.6 the coefficients for these two types of assays.

It should be stressed here that the example we have been analyzing is symmetrical in the sense that the assay has the same number of dose levels for each preparation, each dose is administered the same number of times and the ratios between all dose levels are equal and the same for both preparations (i.e., intervals between doses are equal on the log scale). There is another approach to this testing stage which does not utilize the above contrast concept and in fact the ANOVA table is also more general in the sense that it works for both symmetrical and asymmetrical assays. This is discussed in section 3.14. However, drawbacks to the analysis are the lack of motivation behind the formulas and the untidy calculations.

Table 3.6

THE COEFFICIENTS FOR
THE ORTHOGONAL LINEAR CONTRASTS

The 4-Point Assay

	T_1	T_2	T_3	T_4
L_1	-1	$+1$	-1	$+1$
L_2	$+1$	-1	-1	$+1$
L_3	-1	-1	$+1$	$+1$

The 8-Point Assay

	T_1	T_2	T_3	T_4	T_5	T_6	T_7	T_8
L_1	1	-1	-1	1	1	-1	-1	1
L_2	-1	1	1	-1	1	-1	-1	1
L_3	-1	-1	-1	-1	1	1	1	1
L_4	-3	-1	1	3	-3	-1	1	3
L_5	3	1	-1	-3	-3	-1	1	3
L_6	-1	3	-3	1	-1	3	-3	1
L_7	1	-3	3	-1	-1	3	-3	1

3.4   Point Estimate of ρ

We are now ready for the next stage of the analysis: to estimate the relative potency. For the point estimate we use:

Theorem 3.1   If $M = \log \hat{\rho}$, then

$$M = \frac{\overline{Y}_t - \overline{Y}_s}{\hat{b}} - (\overline{X}_t - \overline{X}_s), \tag{3.6}$$

where $X = \log$ of dose levels, Y is the response and $\hat{b}$ is the estimate of the (common) slope of the parallel lines.

Proof

Let d_t and d_s denote the doses for the test and standard preparations respectively and $X_t = \log d_t$ and $X_s = \log d_s$.

For each fixed response Y, there corresponds two values X_s and X_t (due to parallelism). Since

$$\rho = \frac{1/d_t}{1/d_s} = d_s/d_t$$

then

$$\log \rho = \log d_s - \log d_t = X_s - X_t;$$

and provided the lines are parallel $X_s - X_t$ is the same for each fixed value of Y. Let the lines be written as $Y_t = a_t + b_t X_t$, $Y_s = a_s + b_s X_s$. For each fixed value of Y (then $Y_t = Y_s$) and if the lines are parallel ($b_t = b_s = b$) we must have

$$a_t + b X_t = a_s + b X_s, \text{ i.e., } X_s - X_t = (a_t - a_s)/b.$$

Thus

$$\log \rho = (a_t - a_s)/b;$$

and we take as its estimate

$$M = \log \hat{\rho} = \frac{\hat{a}_t - \hat{a}_s}{\hat{b}}, \tag{3.7}$$

where

$$\hat{a}_t = \overline{Y}_t - \hat{b} \, \overline{X}_t, \quad \hat{a}_s = \overline{Y}_s - \hat{b} \, \overline{X}_s, \tag{3.8}$$

the usual form of the regression estimates of the intercept.

The key problem is to estimate b, the common slope value for both preparations. (We will discuss the estimation of b in the following paragraphs.) Hence substitution of (3.8) into (3.7) yields our result (3.6).

3.5 *Estimation of b*

Since $b_t = b_s = b$ then the best estimate of b is some average of the estimates of b_t and b_s. The usual statistical procedure is to weigh by the reciprocal of the variances of the quantities; i.e.,

$$\hat{b} = \frac{w_s \, \hat{b}_s + w_t \, \hat{b}_t}{w_s + w_t} \tag{3.9}$$

where

$$w_s = \frac{1}{V[\hat{b}_s]} = \frac{\Sigma(X_s - \overline{X}_s)^2}{s^2}$$

$$w_t = \frac{1}{V[\hat{b}_t]} = \frac{\Sigma(X_t - \overline{X}_t)^2}{s^2}$$

where s^2 is the estimated error mean square in the ANOVA table. From the standard regression estimates of $\hat{b}_s$ and $\hat{b}_t$ we finally obtain

$$\hat{b} = \frac{\Sigma(Y_s - \overline{Y}_s)(X_s - \overline{X}_s) + \Sigma(Y_t - \overline{Y}_t)(X_t - \overline{X}_t)}{\Sigma(X_s - \overline{X}_s)^2 + \Sigma(X_t - \overline{X}_t)^2}. \tag{3.10}$$

For symmetrical assays with dose levels varying in the same fashion for both test and standard $w_s = w_t$; and hence in this case

$$\hat{b} = \frac{1}{2}(\hat{b}_s + \hat{b}_t) \ . \tag{3.11}$$

Example: For our example we have from Table 3.1

$$\hat{b} = \frac{1}{2}(10.9624 + 11.4607) = 11.2115 \tag{3.12}$$

The point estimate of ρ for our example can now be found by using (3.6) and (3.11):

$$M = \log \hat{\rho} = \frac{9.80 - 8.17}{11.2115} = 0.14539 \tag{3.13}$$

$$\hat{\rho} = R = \text{antilog (M)} = 1.3976 \tag{3.14}$$

This means that one unit of test is as "potent" as 1.4 units of the standard. (Note $R = 10^M$.)

3.6 *Precision of estimation*

The usual measure of the precision of an estimator is the estimate of the standard deviation of the estimator. In our case this means we have to find the estimate of $(V[M])^{1/2}$, the standard error of M.

The proof of the following result is given in Appendix 5:

Theorem 3.2   The variance of the estimator of the log relative potency is

$$V[M] \doteq \frac{\sigma^2}{b^2}\left\{ \frac{1}{n_t} + \frac{1}{n_s} + \frac{1}{D}(\overline{X}_t - \overline{X}_s + \log \rho)^2 \right\} \ . \tag{3.15}$$

The estimate of this quantity is given by replacing b, σ^2 and $\log \rho$ by their estimates $\hat{b}$, $\hat{\sigma}^2$ and $\log \hat{\rho}$ given in (3.10) or (3.11), (3.17) of the next section, and (3.6) respectively. The quantity D is defined by

$$D = \Sigma(X_s - \overline{X}_s)^2 + \Sigma(X_t - \overline{X}_t)^2 \ . \tag{3.16}$$

Note that if $n_t = n_s = n$ and the assay is symmetric then

$$SE(M) = \left(\frac{MS(E)}{\hat{b}^2} \left(\frac{2}{n} + \frac{M^2}{D} \right) \right)^{1/2} \ .$$

3.7 *Estimation of σ^2*

Since this example utilized a randomized complete block design then the natural estimator is

$$\hat{\sigma}^2 = \text{MS (E)} = 0.15829, \text{ and } S = 0.39785. \tag{3.17}$$

Hence the estimate of the variance of the estimator for the log relative potency, from (3.15), is

$$\hat{V}[M] = \frac{0.15829}{(11.2115)^2} \left\{ \frac{1}{24} + \frac{1}{24} + \frac{(0.145386)^2}{2.8998} \right\} \tag{3.18}$$

$$= 0.0012794.$$

The precision then of our estimate is extremely good, in fact $\sqrt{\hat{V}[M]} \simeq 0.036$. Often we write $M \pm SE(M)$, which in this case is 0.145 ± 0.036.

3.8   *The variance of* $\hat{b}$

Theorem 3.3   Under the above considerations

$$V[\hat{b}] = \sigma^2/D. \tag{3.19}$$

Proof: See Appendix 4.

Therefore $SE(\hat{b}) = S/\sqrt{D}$, and for our example, $SE(\hat{b}) = \left(\dfrac{0.1583}{2.8998} \right)^{1/2}$

$= 0.23$. Thus $\hat{b} \pm SE(\hat{b})$ is 11.21 ± 0.23.

3.9   *A measure of the sensitivity of an assay*

The precision of the estimate of log ρ is strongly influenced by the first factor in (3.13), the ratio σ/b. In fact the sensitivity of an assay is defined by

$$\lambda = \frac{\sigma}{b}, \tag{3.20}$$

and estimated by $\hat{\lambda} = \hat{\sigma}/\hat{b}$. This measure is used to compare assays and was first introduced in Bliss and Cattell (1943).

To see why it is a good measure consider the following arguments. In general for fixed b, the smaller σ is, the smaller the variability and hence the smaller λ is (an indication of a good assay). Notice that if σ is fixed, large values of b make λ small. But large values of b means steep slopes in the regression lines, which means little changes in dose cause large changes in the response; suggesting a very sensitive assay. Similarly if b is small then λ is large; but b small means the dose-response line is not steep; and this means large changes in dose cause little change in response. This indicates a very insensitive assay and hence an undesirable trait. Thus small values of λ indicate sensitivity; large values of λ can be attributed to an insensitive assay.

This is why λ is a good measure of the sensitivity of an assay. But primarily it is used to compare assays. (It is interesting to note that this measure is location and scale invariant.)

In our example $\hat{\lambda} = 0.035$ which is relatively small and indicates a "sensitive" assay.

3.10   Confidence limits for $b_t - b_s$

If we find the confidence limits for the difference of the slopes, i.e., $b_t - b_s$, and if these limits contain zero, we have some evidence for parallelism. It can be shown that the confidence limits for $b_t - b_s$ are defined by

$$(\hat{b}_t - \hat{b}_s) \pm t(v) \sqrt{2\hat{\sigma}^2/\Sigma(X - \overline{X})^2}, \tag{3.21}$$

where $v = $ d.f. of MS(E).

For our data we have the 95% confidence limits

$$11.4607 - 10.9624 \pm 2.03 \sqrt{2(0.158286)/1.4499}$$

$$= 0.4983 \pm 0.9486, \text{ or } (-0.45, +1.45),$$

which contains zero and suggests parallelism. Of course, we have already tested this by the contrast L_2 in our earlier analysis of variance procedure—the results are identical!

3.11   Confidence limits for $\log \rho$

The next stage of the analysis is to find the interval estimate of the log relative potency, $\log \rho$.

Theorem 3.4   The confidence limits for the log relative potency are

$$\frac{M \pm t \dfrac{S}{\hat{b}} \sqrt{\dfrac{M^2}{D} + (1 - g)\left(\dfrac{1}{n_s} + \dfrac{1}{n_t}\right)}}{1 - g}, \tag{3.22}$$

where $M = \log\hat{\rho}$ is given in (3.6), $S = \hat{\sigma}, \hat{b}$ is given in (3.10) or (3.11), D is given in (3.16), t is the Student's t random variable v degrees of freedom and $g = t^2S^2/Db^2$. (Notice that the g in (3.22) is not the same g as in (2.2) and v = df of MS(E).)

Proof: See Appendix 6.

It is important to stress here that (3.22) holds only when $\overline{X}_t = \overline{X}_s$, i.e., when the dosage levels for the standard and test preparations are the same. This is the most common situation in such parallel line assays and usually experiments are designed with this in view. The more general formula, for asymmetrical types, is given in Appendix 6.

Remark: Whenever g is very small (i.e., negligible) and n_s and n_t are very large, these limits can be approximated by

$$M \pm 2 S^* \text{ where } S^* = \frac{SM}{\hat{b} \sqrt{D}}. \tag{3.23}$$

However, this simple formula provides a poor interval estimate and it is not recommended.

In our example, we have $M = 0.1454$, $t = t_{.975}(35) = 2.03$, $\hat{b} = 11.2115$, $n_s = n_t = 24$, $D = 2.8998$, $S^2 = 0.1583$, $g = 0.0003$; thus we obtain from (3.22) the limits (0.1237, 0.1671). So, the interval estimate for ρ (by taking antilogs to the base 10) is (1.33, 1.47), which of course contains the point estimate 1.40.

3.12   *Combination of assays–the weighted mean method*

Suppose our experiment had been repeated k times. Various methods of combining assay results have been proposed in the literature, all involving some element of approximation. (See for example, Armitage et al. (1974) for some references.)

The usual method of combining the estimates has been by a weighted mean of the log relative potencies, the weights being inversely proportional to the estimated variances. Indeed for the ith assay $(i = 1, 2, . . ., k)$ we have (compare with equation (3.6))

$$M_i = \frac{d_i}{\hat{b}_i} - Z_i \tag{3.24}$$

where d_i is the difference $\overline{Y}_t - \overline{Y}_s$ and Z_i is the difference $\overline{X}_t - \overline{X}_s$ for the ith assay, for the estimate of log ρ_i. Then the true log relative potency is estimated by

$$\overline{M} = \frac{\sum_{i=1}^{k} w_i M_i}{\sum_{i=1}^{k} w_i} \tag{3.25}$$

where

$$w_i = \frac{1}{\hat{V}[M_i]}.$$

(Recall that in our notation we mean $\hat{V}[M_i]$ estimates $V[M_i]$.)

$$\hat{V}[M_i] = \frac{\hat{\sigma}^2}{\hat{b}_i^2} \left(n_i^* + \frac{(Z_i + M_i)^2}{D_i} \right) \tag{3.26}$$

where $\hat{\sigma}^2$ is the pooled variance of the k assays, and

$$n_i^* = \frac{1}{n_{t_i}} + \frac{1}{n_{s_i}}.$$

(This is just (3.15) with the subscript i.)

An estimate of the asymptotic variance of $\overline{M}$ can be shown to be simply $1/\sum_{i=1}^{k} w_i$. (See exercise 3.3 and Bennett (1962).)

Recently there has been some research in another method for combining assays: the maximum likelihood approach. (See Armitage (1970).) Although the restrictions are minimal, the method essentially is an iterative procedure requiring a starting value for the estimate of the variance of the common slope b. Armitage et al. (1974), (1976) have shown that when the variability of this slope parameter is negligible, the weighted mean procedure and maximum likelihood yield the same result (an asymptotic or large sample type of equivalence).

Other estimation procedures are also available, for example recently Miller (1973), (1974) has considered the so-called "jackknife" procedure and with respect to two different examples concludes that "the jackknife performed very well . . . in terms of reproducing the Fieller intervals and improving the point estimates". (For those interested in learning about the jackknife procedure, see Brillinger (1964), Gray and Schucany (1972) and Gray, Schucany and Watkins (1975). For a Bayesian approach, see Darby (1980).

3.13   *Slope ratio assays*

In many microbiological assays, the test subject is not an animal but an inoculum (of specified size) of a bacterial culture; the response is some measure of the bacterial growth in a fixed time or of the amount of alkali required to neutralize the acid formed during growth. For such microbiological experiments the slope ratio type of assay is quite common (Finney (1964), p. 188). This section presents a collection of results from Finney (1964), Brown (1964) and Zelen (1970). The latter two are lecture notes which may be inaccessible to most students.

The model assumes that the mean response to the standard preparation is linearly related to some known power of the dose. Then the regression line of Y_s on d_s can be written as

$$E[Y_s | d_s] = a + b_s \, d_s^\lambda.$$

Finney (1964) states "In current use of this type, $\lambda = 1$ appears to be an adequate approximation to the truth." (p. 187)

If the dosage of the test preparation is equivalent to $d_s = \rho\, d_t$ units of the standard preparation, where ρ would be the relative potency, then

$$E[Y_t|d_t] = E[Y_s|\rho\, d_t]$$
$$= a + b_s\,(\rho\, d_t)^\lambda$$
$$= a + b_s\,\rho^\lambda\, d_t^\lambda.$$

If we let the dose metameter be $X = d^\lambda$ and $b_t = b_s\,\rho^\lambda$, then we have

$$E[Y_t|d_t] = a + b_t\, X_t,$$

and

$$E[Y_s|d_s] = a + b_s\, X_s.$$

Therefore the regression of Y_t on X_t is also linear with the same intercept but a different slope. These models would make the response versus dose metameter diagram look like

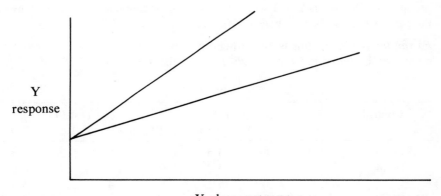

Y
response

X, dose metameter

Since $b_t = b_s\,\rho^\lambda$ then the relative potency of the test to the standard is now defined as

$$\rho = (b_t/b_s)^{1/\lambda}, \tag{3.27}$$

and if $\lambda = 1$, the relative potency becomes the ratio of the slopes, hence the name for such assays. The estimator of ρ is

$$\hat{\rho} = (\hat{b}_t/\hat{b}_s)^{1/\lambda}, \tag{3.28}$$

where $\hat{b}_t$ and $\hat{b}_s$ are the least squares estimates of the corresponding slopes. When

$\lambda = 1$, and this is usually the case, $\hat{\rho} = \hat{b}_t/\hat{b}_s$. The data for slope ratio assays has the following general structure:

Control 0	Standard $X_{s,1}$	$\cdots$	X_{s,k_s}	Test X_{t,k_s+1}	$\cdots$	X_{t,k_s+k_t}
$Y_{0,1}$	$Y_{1,1}$	$\cdots$	$Y_{k_s,1}$	$Y_{k_s+1,1}$	$\cdots$	$Y_{k,1}$
$Y_{0,2}$	$Y_{1,2}$	$\cdots$	$Y_{k_s,2}$	$Y_{k_s+1,2}$	$\cdots$	$Y_{k,2}$
.	.		.	.		.
.	.		.	.		.
.	.		.	.		.
Y_{0,n_0}	Y_{1,n_1}	$\cdots$	$Y_{k_s,n_{k_s}}$	$Y_{k_s+1,n_{k_s}+1}$	$\cdots$	Y_{k,n_k}

Notice that we have included a control or blank dosage. The number of dose groups, or columns, is $k + 1$ where $k = k_s + k_t$; the number of observations in the i-th dose group is n_i, where $i = 0, 1, 2, \ldots, k$.

Example: Although this is an artificial example, references to other experiments can be found in Finney (1964), ch. 7; also see Robel (1978).

Table 3.7

	Control 0	Standard (S) 0.05	0.10	0.15	0.20	Test (T) 1.0	1.5	2.0
	.3	2.1	4.0	5.5	6.6	3.6	5.2	7.0
	.4	2.0	3.7	5.0	6.9	3.8	5.4	6.6
	.5	2.2	3.7	5.1	6.9	3.4	5.3	6.8
$Y_{i.}$	1.2	6.3	11.4	15.6	20.4	10.8	15.9	20.4
$\overline{Y}_i$	0.4	2.1	3.8	5.2	6.8	3.6	5.3	6.8

Before presenting the complete analysis of slope ratio assays it will be necessary to set up the notation. Then, we will find the dose-response relationships (the regression lines for each preparation), the ANOVA table to test for statistical validity, and finally the point and interval estimates for the relative potency. We will use the data of the above examples to explain the notation and analysis.

Notation for Analysis:

$i = 0, 1, 2, \ldots, k$ (there are $k + 1 = k_s + k_t + 1$ dose groups)

n_i = number of responses for i-th dose group

$j = 1, 2, \ldots, n_i$

$$n_s = \sum_{i=1}^{k_s} n_i = \text{total number of responses in S}$$

$$n_t = \sum_{i=k_s+1}^{k} n_i = \text{total number of responses in T}$$

$n = n_0 + n_s + n_t$ = total number of responses

n_{si} = number of responses in S for i-th dose level

(In our example: $k_s = 4$, $k_t = 3$, $k = 7$, $n_{s1} = n_{s2} = n_{s3} = n_{s4} = 3$, $n_{t1} = n_{t2} =$

$n_{t3} = 3$, $n_s = \sum_{i=1}^{4} n_{si} = 12$, $n_t = \sum_{i=5}^{7} n_{ti} = 9$)

X_{si} = i-th dose level for S

$$\overline{X}_s = \frac{1}{n_s} \sum_{i=1}^{k} n_{si} X_{si} = \text{mean dose for S}$$

$$Y_{si.} = \sum_{j=1}^{n_{si}} Y_{sij} = \text{total response for i-th level of S}$$

$$\overline{Y}_{si} = \frac{1}{n_{si}} Y_{si.} = \text{mean response of i-th level of S}$$

$$\overline{Y}_s = \frac{1}{n_s} \sum_{i=1}^{k_s} n_{si} \overline{Y}_{si} = \text{overall mean response for S}$$

Similar quantities can be defined for the T, the test preparation, by replacing s by t.

Table 3.8

i	X_{si}	$Y_{si.}$	$X_{si} \; Y_{si.}$	X_{si}^2
1	.05	6.3	.315	.0025
2	.10	11.4	1.140	.0100
3	.15	15.6	2.340	.0225
$k_s = 4$	.20	20.4	4.080	.0400
Total	0.50	53.7	7.875	0.0750

$$\overline{X}_s = \frac{1}{12}(1.5) = 0.125, \quad \overline{Y}_s = \frac{1}{12}(53.7) = 4.475$$

$$S_X(s, s) = \sum_{i=1}^{k_s} n_{si}(X_{si} - \overline{X}_s)^2 = \text{sum of squares due to dose levels for S.}$$

It is not difficult to show that this reduces to

$$S_X(s, s) = \sum_{i=1}^{k_s} n_{si} \, X_{si}^2 - \frac{1}{n_s}\left(\sum_{i=1}^{k_s} n_{si} \, X_{si} \right)^2$$

and further if $n_{s1} = n_{s2} = \ldots = n_{sk_s}$,

$$S_X(s, s) = n_{s1}\left(\Sigma \, X_{si}^2 - \frac{1}{k_s}(\Sigma \, X_{si})^2 \right). \tag{3.29}$$

For our example: $S_X(s, s) = 3(.0750 - \frac{1}{4}(.50)^2) = 0.0375$.

$$S(Y, s) = \sum_{i=1}^{k_s} n_{si} \, (\overline{Y}_{si} - \overline{Y}_s)(X_{si} - \overline{X}_s) = \text{sum of squares due to responses}$$

and doses in S. It is not difficult to show that this reduces to

$$S(Y, s) = \sum_{i=1}^{k_s} X_{si} \, Y_{si.} - \frac{1}{n_s}(\Sigma \, n_{si} \, X_{si})(\Sigma \, Y_{si.}). \tag{3.30}$$

For our example: $S(Y, s) = 7.875 - \frac{1}{12}(1.5)(53.7) = 1.1625$. The dose-response relation for S is

$$Y'_s = \hat{a}_s + \hat{b}_s \, X_s,$$

where

$$\hat{b}_s = S(Y, s)/S(s, s), \quad \hat{a}_s = \overline{Y}_s - \hat{b}_s \, \overline{X}_s. \tag{3.31}$$

For our example: $\hat{b}_s = 1.1625/.0375 = 31.0$, $\hat{a}_s = 0.6$, $Y'_s = 0.60 + 31 \, X_s$. For the test preparation $k_t = 3$, $n_t = 9$.

Table 3.9

i	X_{ti}	$Y_{ti.}$	$X_{ti} Y_{ti.}$	X_{ti}^2
1	1.0	10.8	10.80	1.00
2	1.5	15.9	23.85	2.25
$k_t = 3$	2.0	20.4	40.80	4.00
Total	4.5	47.1	75.45	7.25

$$\overline{X}_t = \frac{1}{9}(13.5) = 1.5, \; \overline{Y}_t = \frac{1}{9}(47.1) = 5.233$$

$$S_X(t, t) = 3(7.25 - \frac{1}{3}(4.5)^2) = 1.5$$

$$S(Y, t) = 75.45 - \frac{1}{9}(13.5)(47.1) = 4.8$$

$$\hat{b}_t = 4.8/1.5 = 3.2$$

$$\hat{a}_t = 5.233 - (3.2)(1.5) = 0.433$$

$$Y_t' = 0.43 + 3.2 X_t. \tag{3.32}$$

At this stage we should construct a mean response versus dose plot and superimpose these two regression lines on the diagram. We should see how the Y-intercepts relate to the mean response for the control preparation.

ANOVA

In his lectures, Brown (1964) has recommended the following analysis and ANOVA table for slope ratio assays. It has the advantage that it is easy to compute, it yields an estimate of σ^2, it provides tests of the validity of several assumptions, and it is a convenient way to summarize the data.

Four tests are provided by the ANOVA table; the null hypotheses are:

$H_1 : a_s = a_t$ (i.e., equality of intercepts)

$H_2 : a_s = a_t$ = average response for control dose

$H_3 : b_s = b_t = 0$ (i.e., no regression)

$H_4 :$ the regression lines are straight

Table 3.10
GENERAL ANOVA TABLE FOR SLOPE RATIO ASSAYS

Source of variation	df	SS
Among dose groups	$k_s + k_t$	SS(1)
Difference in intercepts	1	SS(2)
Control	1	SS(3)
Regression	2	SS(4)
Linearity	$k_s + k_t - 4$	SS(5)
Within dose groups	$n - k_s - k_t - 1$	SS(E)
Total	$n - 1$	SS(T)

The sum of squares, SS, requires the following quantities, where u denotes s or t:

$$w_u = \sum_{i=1}^{k_u} n_{ui} (X_{ui} - \overline{X}_u)^2 = S_X(u, u). \tag{3.33}$$

$$c_u = \frac{1}{n_u} + \frac{(\overline{X}_u)^2}{w_u}, \quad c = 1/(c_s^{-1} + c_t^{-1}) \tag{3.34}$$

$$SS(1) = \sum_{c,s,t} \sum_{i=1}^{k_u} n_{ui} (\overline{Y}_{ui} - \overline{Y})^2$$

where

$\sum_{c,s,t}$ means sum over control, standard and test groups,

$$SS(2) = (c_s + c_t)^{-1} (\hat{a}_t - \hat{a}_s)^2$$

(Notice that if $\hat{a}_s \simeq \hat{a}_t$, then SS(2) $\simeq 0$.)

$$SS(3) = \left[\frac{1}{n_0} + c \right]^{-1} \left(\overline{Y}_0 - c \left(\frac{\hat{a}_s}{c_s} + \frac{\hat{a}_t}{c_t} \right) \right)^2$$

(Notice that if the mean control dose is near the weighted average of the intercepts, then SS(3) is near zero.)

$$SS(4) = \sum_{u=s,t} w_u \hat{b}_u^2$$

$$SS(5) = \sum_{u=s,t} \sum_{i=1}^{k_u} n_{ui} (\overline{Y}_{ui} - \hat{a}_u - \hat{b}_u X_{ui})^2$$

$$SS(E) = \sum_{c,s,t} \sum_{i=1}^{k_u} \sum_{j=1}^{n_{ui}} (Y_{uij} - \overline{Y}_{ui})^2$$

$$SS(T) = \sum_{c,s,t} \sum_{i=1}^{k_u} \sum_{j=1}^{n_{ui}} (Y_{uij} - \overline{Y})^2.$$

(An APL program for this analysis can be found in Pursey and Hubert (1977).) The four hypotheses are associated with the four sources of variation listed below the between groups variation. It can be shown that these four sources have corresponding sum of squares which are not mutually independent and they do not sum up to the SS between groups. The MS within groups, MS(E), say, is an unbiased estimator of σ^2 and is the denominator for the four F-ratios associated with the four hypotheses, in the sequence given earlier.

The four tests are:

(1) Reject $H_1 : a_s = a_t$ if $F = \dfrac{SS(2)}{MS(E)} > F_{1,v}$.

We hope to accept H_1 because we require that the lines intercept at the right point for fundamental validity of the experiment. (This corresponds to the parallelism in parallel line assays.) If we reject H_1 then the experiment may have to be reanalyzed at lower dose levels where possibly the regression lines have different curvatures.

(2) Reject H_2 : average of intercepts = average response for control group if $F = \dfrac{SS(3)}{MS(E)} > F_{1,v}$, where $v = n - k_s - k_t - 1$.

We hope to accept this hypothesis because for a valid experiment we should have the mean response for the control group (zero or blank dose) not significantly different from the intercepts a_s and a_t. This is a test of whether the underlying model is valid for zero dosage. If we do reject, then there is evidence of statistical invalidity and suggests curvature at very low doses.

(3) Reject $H_3 : b_s = b_t = 0$ if $F = \dfrac{SS(4)/2}{MS(E)} > F_{2,v}$.

We hope to reject H_3 so that we have a significant regression.

(4) Reject H_4 : lines are straight if $F = \dfrac{SS(5)/(k_s + k_t - 4)}{MS(E)} > F_{k_s+k_t-4,v}$.

We hope to accept H_4 because linearity is necessary for statistical validity. If we do reject, an alternative approach may have to be considered.

For our example we have:

$$Y_s' = 0.60 + 31\ X_s, Y_t' = 0.433 + 3.2\ X_t,$$

$$w_s = S_X(s, s) = 0.0375, \quad w_t = S_X(t, t) = 1.50,$$

$$c_s = \frac{1}{12} + \frac{(0.125)^2}{0.0375} = 0.50, \quad c_t = \frac{1}{9} + \frac{(1.5)^2}{1.5} = 1.611,$$

$$c = 1/\left(\frac{1}{.5} + \frac{1}{1.611}\right) = 0.38158$$

$$\overline{Y} = \frac{1}{n} \sum_{c,s,t} \sum_{i=1}^{ku} \sum_{j=1}^{n_{ui}} Y_{uij} = \frac{1}{n} \sum_{c,s,t} \sum_{i=1}^{ku} Y_{ui.}$$

$$= \frac{1}{n}\left(Y_0. + \sum_{i=1}^{k_s} Y_{si.} + \sum_{i=1}^{k_t} Y_{ti.}\right)$$

$$\overline{Y} = \frac{1}{24}(1.2 + 53.7 + 47.1) = 4.25.$$

If all n_i $(i = 0, 1, 2, \ldots, k)$ are equal to n_1, say, then:

$$SS(1) = n_1 \sum_{i=0}^{k} (\overline{Y}_i - \overline{Y})^2 = 3(35.08) = 105.24$$

$SS(T)$ = total SS for all 24 observations = 105.72

$$SS(2) = (.47368)(0.433 - 0.6)^2 = 0.01316$$

$$SS(3) = \left(\frac{1}{3} + 0.38158\right)^{-1}\left(0.4 - 0.38158\left(\frac{0.6}{0.5} + \frac{0.433}{1.611}\right)\right)^2$$

$$= 0.03605$$

$$SS(4) = 0.0375(31)^2 + 1.5(3.2)^2 = 51.3975$$

$$SS(5) = 3(0.015 + 0.0066) = 0.0650$$

$$SS(E) = 0.02 + 0.28 + 0.18 = 0.48.$$

Table 3.11
ANOVA TABLE FOR EXAMPLE

Source	d.f.	SS	MS	F	$F_{.95}$
Among groups	7	105.2400	—	—	—
Control	1	.01316	.01316	0.44	4.5
Intercepts	1	.03605	.03605	1.20	4.5
Regression	2	51.3975	25.69875	856.6	3.6
Linearity	3	.0650	.02167	0.72	3.25
Within groups	16	.4800	.03	—	—
Total	23	105.72	—	—	—

We see that (1) there is statistical validity (i.e., the mean response of control group is not significantly different from the intercepts), (2) there is fundamental validity (i.e., there is no significant difference between the two intercepts), (3) there is a significant regression, and (4) the lines are straight.

We can see here that the partitioning of the among groups SS into the four other sources of variation is such that the SS do not add up! This is due to the fact that these four sources have sums of squares which are not mutually independent (Brown, 1964). Indeed, one usually finds $\Sigma_{j=2}^{5} SS(j) < SS(1)$. However, if one approaches the analysis as a multiple regression problem (see our estimation treatment below) where the model for the response variable is

$$E[Y_{ij}] = a + b_s X_{si} + b_t X_{ti}, i = 0, 1, 2, \ldots, k ; j = 1, 2, \ldots, n_i \quad (3.35)$$

$$V[Y_{ij}] = \sigma^2, \text{ for all } i, j,$$

Y_{ij} are mutually independent normal variables,

then it is possible to achieve an ANOVA table where the partition of the sums of squares is "orthogonal" (see Finney (1964), page 196). By the way, the normality assumption is not crucial because this method of approach is robust and asymptotically exact even without this assumption. (See Finney (1951a), (1964), Claringbold (1959).)

Estimation of parameters

Provided the testing stage of the analysis has been satisfactory (such as in our example) we are ready to estimate the parameters. Estimation, like ANOVA, requires a statistical framework and a method of estimation; we shall use the method of least squares in the multiple regression setting.

The data can be recast into the following form:

Table 3.12

Dose group	Independent Variables X_s	X_t	Dependent Variables
0	0	0	$Y_{01}, Y_{02}, \ldots, Y_{0n_0}$
1	X_{s1}	0	$Y_{11}, Y_{12}, \ldots, Y_{1n_1}$
2	X_{s2}	0	$Y_{21}, Y_{22}, \ldots, Y_{2n_2}$
.	.	.	.
.	.	.	.
.	.	.	.
k_s	X_{sk_s}	0	$Y_{k_s1}, Y_{k_s2}, \ldots, Y_{k_sn_{k_s}}$
$k_s + 1$	0	X_{t,k_s+1}	$Y_{k_s+1,1}, Y_{k_s+1,2}, \ldots,$ $Y_{k_s+1,n_{k_s+1}}$
$k_s + 2$	0	X_{t,k_s+2}	.
.	.	.	.
.	.	.	.
.	.	.	.
k	0	$X_{t,k}$	$Y_{k1}, Y_{k2}, \ldots, Y_{kn_k}$

In the multiple regression setting, the observations are triplets (X_s, X_t, Y) and there are $n = \sum_{i=0}^{k} n_i$ such triplets. (For our example there are 24 such triplets.) Because of this setting we use a slightly different notation; let:

$$X_{u.} = \sum_{i=0}^{k} n_i X_{ui}, \ u = \text{s or t.}$$

(Notice that $X_{s.}$ is the sum of the X_s values for all n triplets; it is not simply the sum of the standard dose levels.)

$$\overline{X}_{u.} = \frac{1}{n} X_{u.}$$

$$Y_{i.} = \sum_{j=1}^{n_i} Y_{ij} = \text{i-th dose group total}$$

$$\overline{Y}_{i.} = \frac{1}{n_i} Y_{i.} = \text{mean of i-th dose group}$$

$$Y.. = \sum_{i=0}^{k} \sum_{j=1}^{n_i} Y_{ij} = \sum_{i=0}^{k} Y_{i.} = \text{total of all responses}$$

$$\overline{Y}.. = \frac{1}{n} Y.. = \text{mean of Y observations.}$$

$$S_{yy} = \sum_{i=0}^{k} \sum_{j=1}^{n_i} (Y_{ij} - \overline{Y}..)^2 = \sum_{i=0}^{k} \sum_{j=1}^{n_i} Y_{ij}^2 - \frac{1}{n} Y_{..}^2, \tag{3.37}$$

which is SS(T) in the ANOVA.

$$S_{yu} = \sum_{i=0}^{k} \sum_{j=1}^{n_i} (Y_{ij} - \overline{Y}..)(X_{ui} - \overline{X}_{u.}) = \sum_{i=0}^{k} X_{ui} Y_{i.}$$

$$- \frac{1}{n} Y.. Y_{u.} \tag{3.38}$$

$$S_{uv} = \sum_{i=0}^{k} n_i (X_{ui} - \overline{X}_{u.})(X_{vi} - \overline{X}_{v.}) = \sum_{i=0}^{k} n_i X_{ui} X_{vi}$$

$$- \frac{1}{n} X_{u.} X_{v.} \tag{3.39}$$

$$S_{uu} = \sum_{i=0}^{k} n_i X_{ui}^2 - \frac{1}{n} X_{u.}^2 \tag{3.40}$$

$$D = S_{ss} S_{tt} - S_{st}^2 . \tag{3.41}$$

Then we can express the point estimates of the four key parameters as:

$$\hat{b}_s = (S_{ys} S_{tt} - S_{yt} S_{st})/D \tag{3.42}$$

$$\hat{b}_t = (S_{yt} S_{ss} - S_{ys} S_{st})/D \tag{3.43}$$

$$\hat{a} = \overline{Y}.. - \hat{b}_s \overline{X}_{s.} - \hat{b}_t \overline{X}_{t.} \tag{3.44}$$

$$\hat{\sigma}^2 = \frac{1}{n-3} \sum_i \sum_j (Y_{ij} - \hat{a} - \hat{b}_s X_{si} - \hat{b}_t X_{ti})^2 \tag{3.45}$$

$$= \frac{1}{n-3} (S_{yy} - \hat{b}_s S_{ys} - \hat{b}_t S_{yt}).$$

The variance and covariance of the estimators $\hat{b}_s$ and $\hat{b}_t$ are:

$$V[\hat{b}_s] = \sigma^2 S_{tt}/D \tag{3.46}$$

$$V[\hat{b}_t] = \sigma^2 S_{ss}/D \tag{3.47}$$

$$\text{Cov}[\hat{b}_s, \hat{b}_t] = - \sigma^2 S_{st}/D. \tag{3.48}$$

The estimator of the relative potency ρ is:

$$\hat{\rho} = (\hat{b}_t/\hat{b}_s)^{1/\lambda}. \tag{3.49}$$

(Usually $\lambda = 1$.) Since $\hat{\rho}$ is a ratio of random variables, we can use Fieller's theorem (Appendix 1) to derive the confidence limits for ρ^λ. After one obtains these limits, the confidence limits for ρ are found by taking the $1/\lambda$ roots of these limits.

The $(1 - \alpha) \cdot 100\%$ confidence limits for ρ^λ are:

$$\frac{1}{1 - g} \left\{ \hat{\rho} + \frac{g \, S_{st}}{S_{tt}} \pm \frac{t \, \hat{\sigma}}{\hat{b}_s} \right.$$
$$\left. \sqrt{\frac{1}{D} \left[(1 - g)S_{ss} + 2 \, S_{st} \, \hat{\rho} + S_{tt} \, \hat{\rho}^2 + g \, \frac{S_{st}^2}{S_{tt}} \right]} \right\} \tag{3.50}$$

where $t = t_{1-(\alpha/2)} (n - 3)$ and $g = t^2 \, \hat{\sigma}^2 \, S_{tt}/\hat{b}_s^2 \, D$.

If $g \simeq 0$, we have approximately

$$\hat{\rho} \pm \frac{t \, \hat{\sigma}}{\hat{b}_s} \sqrt{\frac{1}{D} (S_{ss} + 2 \, S_{st} \, \hat{\rho} + S_{tt} \, \hat{\rho}^2)}. \tag{3.51}$$

For more comments on this type of assay please see Finney (1945), (1946), (1951a), (1964), Claringbold (1959) and for combining several slope ratio assays (i.e., repeated experiments) see Bennett (1963).

For our example we have:

$$X_{s.} = 1.5, \, X_{t.} = 13.5, \, \overline{X}_{s.} = 0.0625, \, \overline{X}_{t.} = 0.5625, \, Y_{..} = 102.0$$

$$\overline{Y}_{..} = 4.25, n = 24, S_{yy} = 539.22 - \frac{1}{24} (102)^2 = 105.72$$

$$S_{ys} = \sum_{i=0}^{k} X_{si} \, Y_{i.} - \frac{1}{n} Y_{..} \, X_{s.} = 7.875 - \frac{1}{24} (102) (1.5) = 1.5$$

$$S_{yt} = \sum_{i=0}^{k} X_{ti} \, Y_{i.} - \frac{1}{n} Y_{..} \, X_{t.} = 75.45 - \frac{1}{24} (102) (13.5) = 18.075$$

$$S_{ss} = \sum_{i=0}^{k} n_i \, X_{si}^2 - \frac{1}{n} X_{s.}^2 = 0.225 - \frac{1}{24}(1.5)^2 = 0.13125$$

$$S_{tt} = \sum_{i=0}^{k} n_i \, X_{ti}^2 - \frac{1}{n} X_{t.}^2 = 21.75 - \frac{1}{24}(13.5)^2 = 14.15625$$

$$S_{st} = \sum_{i=0}^{k} n_i \ X_{si} \ X_{ti} - \frac{1}{n} X_{s.} \ X_{t.} = 0 - \frac{1}{24}(1.5) \ (13.5) = -0.84375$$

$D = (0.13125) \ (14.15625) - (-0.84375)^2 = 1.14609.$

The least squares estimates of b_s and b_t:

$\hat{b}_s = (1.5 \ (14.15625) - 18.075 \ (-0.84375))/D = 31.8344$

$\hat{b}_t = (18.075 \ (0.13125) - 1.5 \ (-0.84375))/D = 3.17423.$

The least squares estimate of a:

$\hat{a} = 4.25 - 31.8344 \ (0.0625) - 3.17423 \ (0.5625) = 0.47485.$

The multiple regression equation:

$Y' = 0.475 + 31.8 \ X_s + 3.17 \ X_t.$

The estimate of the residual or error variance:

$$\hat{\sigma}^2 = \frac{1}{21} \ (105.72 - 31.8344 \ (1.5) - 3.17423 \ (18.075)) = 0.02829.$$

The estimate of the variance and the standard error of $\hat{b}_s$ and $\hat{b}_t$:

$\hat{V}[\hat{b}_s] = 0.02829 \ (14.15625)/D = 0.34943, \ SE(\hat{b}_s) = 0.59.$

$\hat{V}[\hat{b}_t] = 0.02829 \ (0.13125)/D = 0.00324, \ SE(\hat{b}_t) = 0.057.$

The estimate of the covariance:

$\hat{Cov}[\hat{b}_s, \hat{b}_t] = -0.02829 \ (-0.84375)/D = 0.02083.$

The estimate of the relative potency:

$$\hat{\rho} = \frac{3.17423}{31.8344} = 0.0997 \approx 0.1.$$

The test preparation is only 1/10 as potent as the standard. The 95% confidence limits for ρ are (0.0963, 0.1029).

   If we let g = 0 (in fact, g = 0.0015) the approximate 95% confidence limits for ρ are: $0.0997 \pm .0033$, or (.0964, .1030). (If we had H : $\rho = 1$ vs A : $\rho \neq 1$, then we would reject H and say that the test preparation is significantly less potent than the standard preparation.)

   An APL program called SRAY which performs the statistical analysis for slope ratio assays is given in Pursey and Hubert (1977).

3.14   *On asymmetric parallel-line assays*

The following data was taken from an August 22, 1975 seminar by D. J. Finney in Sherbrooke, Que. It is a 1940 bioassay of vitamin D3 in cod-liver oil done by the British Standards Institution and represents an interesting example of an asymmetric completely randomized parallel-line assay. It is called asymmetric because the number of dose levels for each preparation is not the same, the number of administrations of the dose levels are not the same, and the intervals between the dose levels are not equal on the log scale. However, one can still construct a parallel-line type of diagram of response versus log dose for each preparation on the same sheet of paper. If we let the cod-liver oil be the test preparation then we want to estimate the relative potency of cod-liver oil.

Table 3.13

	Vitamin D3			cod-liver oil (test)			
Dose levels	5.76	9.60	16.00	32.4	54.0	90.0	150.0
Log dose	.7604	.9823	1.2041	1.5105	1.7324	1.9542	2.1761
	33.5	36.2	41.6	32.0	32.6	35.7	44.0
	37.3	35.6	37.9	33.9	37.7	42.8	43.3
	33.0	36.7	40.5	30.2	36.0	38.9	38.4
	33.1	34.8	42.0	33.1	34.8	38.9	40.1
Responses	32.4	39.5	39.1	31.6	29.2	40.3	44.2
	32.1	37.0	42.4	32.7	34.6	38.6	41.8
	33.7	36.2	39.4	28.8		42.9	43.7
	29.5	39.4	43.0			43.9	
	32.8	35.4					
		34.2					
n	9	10	8	7	6	8	7
ΣY	297.4	365.0	325.9	222.3	204.9	322.0	295.5
$\overline{Y}$	33.04	36.50	40.74	31.76	34.15	40.25	42.21
SS	32.682	28.080	22.799	18.537	43.555	53.920	29.709

Because this is not a symmetric type of assay the formulas used in the ANOVA and estimation have to be slightly altered; indeed, for this problem we use the following analysis:

Let Y_{uij} represent the response to the u-th preparation (u is either s or t) at the i-th dose level for that preparation ($i = 1, 2, \ldots, k_u$) and the j-th subject at that level for that preparation ($j = 1, 2, \ldots, n_{ui}$). For Table 3.13: $k_s = 3$, $k_t = 4$ and $n_{s1} = 9$, $n_{s2} = 10$, $n_{s3} = 8$ and $n_{t1} = 7$, $n_{t2} = 6$, $n_{t3} = 8$, $n_{t4} = 7$. Also let $n_u = \Sigma_{i=1}^{k_u} n_{ui}$; then $n_s = 27$ and $n_t = 28$, the total number of subjects given the u-th preparation.

The total mean response for each preparation is defined by

$$\overline{Y}_u = \frac{1}{n_u} \sum_{i=1}^{k_u} \sum_{j=1}^{n_{ui}} Y_{uij}. \tag{3.52}$$

Here,

$$\overline{Y}_s = \frac{1}{27}(297.4 + 365.0 + 325.9) = 36.60$$

$$\overline{Y}_t = \frac{1}{28}(222.3 + 204.9 + 322.0 + 295.5) = 37.31.$$

Let X_{ui} represent the i-th level of the dose metameter at which the u-th preparation is applied. The mean log dose level used for the n_u subjects is

$$\overline{X}_u = \frac{1}{n_u} \sum_{i=1}^{k_u} n_{ui} X_{ui}. \tag{3.53}$$

Here,

$$\overline{X}_s = \frac{1}{27}(9(.7604) + 10(.9833) + 8(1.2041)) = 0.97405$$
$$\overline{X}_t = 1.85124.$$

Provided the assumptions for the analysis are satisfied (see the ANOVA below), the least squares estimates of a_s, a_t and b are given by

$$\hat{a}_s = \overline{Y}_s - \hat{b}\,\overline{X}_s, \quad \hat{a}_t = \overline{Y}_t - \hat{b}\overline{X}_t \tag{3.54}$$

$$\hat{b} = \frac{w_s\,\hat{b}_s + w_t\,\hat{b}_t}{w_s + w_t} \tag{3.55}$$

where

$$w_u = \sum_{i=1}^{k_u} n_{ui}(X_{ui} - \overline{X}_u)^2$$

$$\hat{b}_u = \frac{\displaystyle\sum_{i=1}^{k_u} \sum_{j=1}^{n_{ui}} (X_{ui} - \overline{X}_u)(Y_{uij} - \overline{Y}_u)}{\displaystyle\sum_{i=1}^{k_u} n_{ui}(X_{ui} - \overline{X}_u)^2}.$$

Then the estimate of the log relative potency is $M = (\hat{a}_t - \hat{a}_s)/\hat{b}$.

The general ANOVA table for testing the assumptions for such bioassays appears in Table 3.14. (A justification for this analysis can be found in Brown

(1964).) It is general in the sense that it works for both symmetrical and asymmetrical parallel-line assays. The various sums of squares (see Appendix 7 for proof of the partitioning) are defined by

$$SS(T) = \sum_{u=s}^{t} \sum_{i=1}^{k_u} \sum_{j=1}^{n_{ui}} (Y_{uij} - \overline{Y})^2$$

$$SS(1) = \sum_{u=s}^{t} \sum_{i=1}^{k_u} n_{ui} (\overline{Y}_{ui} - \overline{Y})^2$$

$$SS(E) = \sum_{u=s}^{t} \sum_{i=1}^{k_u} \sum_{j=1}^{n_{ui}} (Y_{uij} - \overline{Y}_{ui})^2$$

where

$SS(1) + SS(E) = SS(T)$. Let $\hat{a}_u = \overline{Y}_u - \hat{b}_u \overline{X}_u \ (u = s, t)$,

$$SS(2) = \sum_{u=s}^{t} \sum_{i=1}^{k_u} n_{ui} (\overline{Y}_{ui} - \hat{a}_u - \hat{b}_u X_{ui})^2$$

$$SS(3) = \sum_{u=s}^{t} w_u (\hat{b}_u - \hat{b})^2$$

$$SS(4) = \sum_{u=s}^{t} n_u (\overline{Y}_u - \overline{Y})^2$$

$$SS(5) = (w_s + w_t) \hat{b}^2$$

where $SS(2) + SS(3) + SS(4) + SS(5) = SS(1)$. Also $MS = SS/df$ and $\overline{Y}$ is the mean of all $n_s + n_t$ observations.

Table 3.14
GENERAL PARALLEL–LINE ASSAY ANOVA

Source of variation	d.f.	SS
Between groups	$k_s + k_t - 1$	SS(1)
Deviations from linearity	$k_s + k_t - 4$	SS(2)
Parallelism	1	SS(3)
Preparations	1	SS(4)
Regression	1	SS(5)
Within groups (error)	$n_s + n_t - k_s - k_t$	SS(E)
Total	$n_s + n_t - 1$	SS(T)

There are 5 tests of hypotheses which can be obtained from this ANOVA table and it is recommended that they be performed in the order given here.

1. To test H : $a_s = a_t$ and $b_s = b_t = 0$.

   Rule: reject H if $F = MS(1)/MS(E) > F_c$.

As usual, F_c is the 95th percentile point with the appropriate degrees of freedom governed by the test statistic. If we accept H then none of the mean response levels for either dose group were significantly different (i.e., all were close to $\overline{Y}$). Therefore the dosage levels used in the experiment were probably all too high or all too low to cause any observable change in the response for either of the preparations. The assay must be redone with different dose levels in this case. Of course, we expect to reject this hypothesis; and if we do, then we proceed with the next test.

2. To test H : the regressions of Y on X for each preparation are linear.

   Rule: reject H if $F = MS(2)/MS(E) > F_c$.

If we reject this hypothesis then the linearity assumption is invalid and the estimation is invalid. Possibly some non-linear (e.g. parabolic) assay methods are applicable. This test can only be done if $k_s + k_t > 4$. We expect to accept H and then proceed on to the next test.

3. To test H : $b_s = b_t = b$ (parallelism).

   Rule: reject H if $F = MS(3)/MS(E) > F_c$.

If we reject H then the assumption of a constant relative potency, ρ, is invalid and the potency of the test preparation relative to the standard depends on the dose level. If we accept H we have our parallelism and proceed on.

4. To test H : $a_s = a_t$.

   Rule: reject H if $F = MS(4)/MS(E) > F_c$.

If we reject this hypothesis then ρ is significantly different from 1 ($\log \rho \neq 0$). If we accept H then ρ is not significantly different than 1 and the potency of the test preparation is approximately that of the standard.

5. To test H : $b = 0$.

   Rule: reject H if $F = MS(5)/MS(E) > F_c$.

If we reject H then there is a significant regression. If we accept H then the dose levels may have been inappropriately chosen since there is no significant change in response across the different dose levels of either preparation.

A general APL program for such experiments can be found in Seth and Hubert (1977).

3.15   *Exercises*

3.1   Since $L_i = \sum_{j=1}^{J} \alpha_{ij} T_j$ then by independence,

$$V[L_i] = \sum_{j=1}^{J} \alpha_{ij}^2 V[T_j] = \sum_{j=1}^{J} \alpha_{ij}^2 n\sigma^2 = n\sigma^2 \sum_{j=1}^{J} \alpha_{ij}^2 .$$

What is the estimator of σ^2?

3.2   The following data show particular values related to a measure of performance of trout in response to fluoride, and to a solution of fluoride preparation containing various impurities as well as an unknown amount of fluoride.

Dose level (ppm)	Standard fluoride dose			Unknown (test) dose			
	2	4	8	2	4	8	Total
	10.5	17.0	28.0	8.5	20.0	25.0	109.0
	8.5	21.5	34.0	5.5	14.0	24.5	108.0
Responses	10.0	16.0	25.5	3.0	16.0	28.0	98.5
	8.0	13.5	31.5	3.0	16.0	24.0	96.0
	10.0	15.0	28.5	2.0	15.0	22.0	92.5
Total	47.0	83.0	147.5	22.0	81.0	123.5	504.0

Five replicates of each of the six doses were given, all to the same tissue. The doses were arranged into five random blocks (the order in which the six doses were arranged in each block was decided strictly at random using random number tables).

What conclusions can be reached? (Hint: your analysis should also include a response diagram, confidence limits for ρ and any other informative measures that may shed light on this data. R = 0.79.)

3.3   In section 3.12 the weighted mean method was described for combining assays. Suppose that the example 3.1 was repeated a total of 4 times. Furthermore suppose that in addition to the data given in Table 3.1; we have the following 3 sets of data. Estimate the true log relative potency by combining the four assays using this weighted mean method.

	Standard			Test		
	.25	.50	1.00	.25	.50	1.00
#2	5.0	8.0	11.0	6.6	9.8	12.4
	4.4	8.0	11.0	6.8	9.6	12.8
	5.0	8.0	10.6	6.6	9.6	12.6
	5.0	8.0	11.0	6.8	9.4	11.8
	5.2	7.6	10.4	7.0	9.6	12.2
	5.0	8.0	10.8	6.8	9.2	12.6
	5.8	8.0	11.0	6.4	9.8	12.2
	5.4	8.0	11.2	6.4	9.6	12.8
#3	5.0	8.0	10.2	6.8	9.8	11.8
	5.0	8.0	10.0	6.8	9.4	11.8
	5.0	7.2	10.8	6.6	8.8	11.8
	4.2	7.0	10.0	6.6	9.2	11.8
	5.0	8.0	10.0	6.0	9.4	11.8
	5.0	8.0	10.2	6.8	9.6	11.8
	5.0	7.4	10.0	6.6	9.0	11.6
	5.0	7.6	10.0	6.4	9.2	11.6
#4	4.2	7.6	9.8	6.0	9.4	11.8
	5.0	8.0	10.2	6.4	9.4	11.8
	4.4	7.2	10.0	6.8	8.8	11.8
	5.0	7.0	10.0	7.0	8.8	11.8
	4.6	7.2	10.0	6.8	9.2	11.8
	4.4	7.0	10.0	6.2	8.8	11.2
	5.0	7.2	10.0	6.4	9.0	11.8
	4.4	7.0	10.0	6.6	9.0	11.8

3.4   In a bioassay of insulin from the blood sugar response of rabbits, each of 4 rabbits was injected with 4 doses, two of the standard at 0.6 and 1.2 units (S_1 = A and S_2 = B) and two of the test preparation at the same levels (U_1 = C, U_2 = D).

The four rabbits were injected on each of 4 days (rows) with the doses A to D of the Latin Square presented below. The response variable Y is the milligram percentage of blood sugar in a 1-ml. sample taken from an ear vein 50 minutes after injection of the insulin. (This assay is discussed in detail in Young and Roman (1948); also see Bliss and Marks (1939).)

Date April	Dose and Y in rabbit number 1	2	3	4	Totals
23	B 24	C 46	D 34	A 48	152
25	D 33	A 58	B 57	C 60	208
26	A 57	D 26	C 60	B 45	188
27	C 46	B 34	A 61	D 47	188
Totals	160	164	212	200	736
ΣX^2	7030	7312	11726	10138	36206

(a) Carry out a complete ANOVA with interpretations.

(b) Estimate the log relative potency and its standard error.

(c) Find the 95% confidence limits for ρ.

(d) What assumptions have been made in this assay?

(e) What conclusions can be drawn from this assay?

3.5 Kempthorne (1952), Section 18.8, suggests that when one factor has a level with zero dosage (control), certain contrasts will include a comparison of the control plots which contribute only to experimental error. Find the implications of this fact. HINT: also read Yates (1937) and Addelman (1974).

3.6 The following data are from a classical assay of a test preparation of testosterone propionate against a standard, using three doses of each. Each of the 6 doses was injected into 5 capons and the increase in the sum of the length and height of the comb was used for assay purposes.

Dose (μg.)	Standard preparation			Test preparation		
	20	40	80	20	40	80
Responses	6	12	19	6	12	16
	6	11	14	6	11	18
	5	12	14	6	12	19
	6	10	15	7	12	16
	7	7	14	4	10	15
Totals ΣY	30	52	76	29	57	84
ΣY^2	182	558	1174	173	653	1422

Set up a complete ANOVA for this experiment, estimate the relative potency of the test preparation and find the associated 95% confidence limits.

STANDARD

0.1	0.2	0.4	0.8	1.6	3.2	6.4	12.8
37	27	56	102	151	194	154	199
15	26	46	68	121	185	200	215
22	32	17	61	102	145	208	165
17	26	24	63	121	177	112	205
31	18	25	68	124	180	136	125
25	37	42	101	114	171	230	177
16	27	27	58	137	202	186	206
18	20	30	81	151	136	160	205
26	18	26	97	125	186	214	210
29	25	39	79	105	93	153	174

TEST

0.07	0.14	0.28	0.56	1.12	2.24	4.48	8.96
19	33	28	44	64	143	204	212
18	19	42	58	63	121	202	208
11	24	20	37	50	132	150	170
18	15	28	32	44	137	145	198
22	19	22	25	83	181	134	206
26	24	31	27	92	150	181	216
20	22	31	42	62	118	162	189
12	24	33	75	83	151	176	192
18	24	19	37	86	120	207	198
28	37	14	47	78	186	181	209

3.7   Two biologists X and Y performed independent comparative bioassays on the effect of a toxic substance on a type of coral from Australia. Each had a test preparation and a standard preparation, and each used three different dosage levels for each preparation. After their experimental results were obtained they both estimated the average slope of the parallel lines to be about 7.5. However, X found the estimate of σ to be 5.0 and Y found it to be 2.5. Which assay was more sensitive and why?

3.8   Mantel and Schneiderman (1975) illustrate a new procedure by an experiment in which the ventral prostate weights (in mg) of male Sprague-Dawley rats are treated with testosterone. The data above is slightly abridged from their data. The total dosages of testosterone shown were administered over a 10-day period, after which the rats were sacrificed and measurements made. The total of 16 dose levels used can be considered as a standard preparation series beginning with 0.1 mg per rat and increasing by multiples of 2 to 12.8 mg/rat together with a test preparation series increasing from 0.07 mg/rat to 8.96 mg/rat. The true relative potency should be 70%. From this data find the point estimate of the relative potency.

3.9   Cohen and Leppink (1956) obtained four independent estimates of the log relative potency, M, of a pertussis vaccine relative to a standard NIH reference vaccine: -1.733, -1.488, -1.831, -1.875, with the estimated variances of these log potencies as 0.04815, 0.02466, 0.03142, and 0.07641, respectively. (This data is analyzed in detail in Finney (1971), chapter 6.) Show that the mean value of M is $\overline{M} = -1.686$ with the asymptotic standard deviation as 0.1.

3.10 The following fictitious data are similar to those reported in Brown (1964) involving an assay on ACTH. It depends on the increased secretion of steroids by the adrenals of the hypophysectomized rat when ACTH is injected. A needle is inserted into the adrenal vein and a dose of ACTH is released so that it backs into the organ. The volume injected is about 1 mℓ. The needle remains in the vein and 90 seconds later 0.6 mℓ of adrenal vein blood is removed. The time required to collect the blood is recorded. The blood is measured for steroids B, and the measurement is expressed in mg B per 100 mℓ plasma per min. per 100 gm body weight.

Standard Preparation (mℓ) $\times 10^3$			Plasma (mℓ) $\times 10$		
5	50	500	5	10	15
3	27	44	17	21	25
6	19	41	6	12	21
4	28	35	11	21	16
11	14	36	14	18	26

We want to estimate the relative potency:

(a) Using semi-logarithmic paper plot the 6 mean responses versus log dose. What does your diagram suggest?

(b) Assuming a CRD do a complete analysis of variance. What can you conclude about the slopes?

(c) Find $\hat{b}$ using the weighted average.

(d) Find M, the point estimate of the log relative potency.

(e) Estimate σ.

(f) Find the standard error of $\hat{b}$.

(g) Find the standard error of M.

(h) Estimate λ.

(i) Find the 95% confidence limits for the relative potency.

3.11 Complete the analysis and find the estimate of the log relative potency for the data of exercise 3.2 according to the general method discussed in section 3.14. (Note: your ANOVA table will have a block source of variation.) To find the 95% confidence limits use the result in Appendix 6 with $S = [MS(error)]^{1/2}$.

3.12 The description of an experiment which would yield data such as the following can be found in Finney (1964, p. 189). Plot a mean response versus dosage diagram. Analyze the data using the slope ratio approach with $\lambda = 1$. What is wrong with this data? (Notice that $X = d$.)

Control	Standard					Test		
(Dosage)	.05	.10	.15	.20	.25	.14	.21	.29
1.5	3.5	5.0	6.2	8.0	9.4	4.9	6.3	7.7
1.4	3.2	4.7	6.1	7.7	9.5	4.8	6.5	7.7

4 Quantal Assays: Single Agent

4.1 *Introduction*

Quantal response assays belong to the class of qualitative indirect bioassays. They are characterized by experiments in which a stimulus (e.g., dose of a drug) is applied to n experimental units and r of them respond and $n - r$ do not respond. An example is given in section 4.2 and the analysis is given in the remaining sections.

One of the objectives of such assays is to estimate what level of the stimulus is necessary to bring about a response in a given percentage of individuals in the population. For any one individual there is a level of intensity of the stimulus below which response does not occur. This level is in some sense the tolerance of the individual; often the amount of stimulus needed to just produce a particular response in a subject is referred to as the *individual effective dose* (IED). Clearly, when we want to estimate the tolerance for a given population of individuals a sample is necessary. The level of a stimulus which will result in a response by 50% of individuals in the population under study is an important characterizing parameter and it is denoted by LD50 for *median lethal dose* (or ED50 for *median effective dose,* LC50 for *median lethal concentration,* EC50 for *median effective concentration,* and Tlm for *median tolerance limit*). By the way, Sprague (1969) has pointed out that the symbol LD50 was first proposed by Trevan (1927) who wrote the symbols all on the same line (not LD_{50}), and further that "making 50 into a subscript is historically incorrect and seems to serve little purpose beyond complicating the lives of typists". Often the time period will be specified by writing, for example, 48-h LC50; the 48 hours labels the length of exposure to the stimulus. As will be seen later, the estimates of the LD50's for preparations can be compared and used to estimate the relative potency. The physical components of such experiments are extremely important and laboratory methods are described in Hoskins and Craig (1962), Bucher and Morse (1963), Sprague (1973) and Banki (1978).

These parameters can also be used to find *thresholds,* below which the concentration of a toxic substance can be considered *safe.* It must be stressed here that since an LC50 is merely a reference point for expressing *acute lethal toxicity* of a given agent, then the safe concentration which allows growth and other processes to continue is lower than LC50. When a drug is *acutely* lethal it causes severe and rapid damage to the subject, usually within a short period of time (e.g., 4 days for some fish); *sublethal* experiments involve levels below which death directly results; a *subacute* experiment involves levels below acute and the responses take longer and even may continue for a long period of time. For a deeper discussion of these terms, see Sprague (1969), (1970), (1971), and Dorfman and Cirello (1978). For other areas of applicability of quantal experiments see Brown (1978).

4.2   *A typical toxicity (quantal) experiment*

The following data show the effect of different concentrations of a toxic material (nicotine sulphate in a 1% saponin solution) on a certain insect (*Drosophila melanogaster*).

Table 4.1
THE FRUIT FLY EXAMPLE

Dose (gm/100cc) d	Number of Insects n	Number Killed r	Percent Killed p
0.10	47	8	17.0
0.15	53	14	26.4
0.20	55	24	43.6
0.30	52	32	61.5
0.50	46	38	82.6
0.70	54	50	92.6
0.95	52	50	96.2

What does this data tell us? Obviously, a fixed dose, d, of the drug is given to a group of n subjects and the number, r, of subjects responding is observed. The percentage of subjects responding in the group is $p = \frac{r}{n} \times 100$. As the dosage is increased the percentage killed increases. Here the dose level is fixed by the experimenter and the number of subjects responding is the variable observed. (This is quite different from a direct assay where the dose is not fixed and is in fact the variable observed by the experimenter.) What type of random variable is r?

Let us concentrate on p. This variable represents the percentage of subjects who individually each have an effective dose equal to or less than d. Recall that the individual effective dose, IED, is the quantity of a drug needed to just produce a specific response; viz., death. For example, in this data p = 17.0 represents the percentage of insects who responded to a dose of d equal to 0.10 units *or* to less than 0.10 units. The point is that the value d = 0.10 is like an upper end point of an interval, i.e., 0 to 0.10, and the p value of 17.0 is the accumulated relative frequency of responses for this interval.

Similarly p = 26.4 represents the percentage of insects who responded to a dose level of d equal to 0.15 units or to less than 0.15 units. That is, the value d = 0.15 can be considered as an upper end point of an interval, viz., (0 − 0.15) and the p value 26.4 represents the accumulated relative frequency of responses for this interval.

A pictorial representation of the data is as follows:

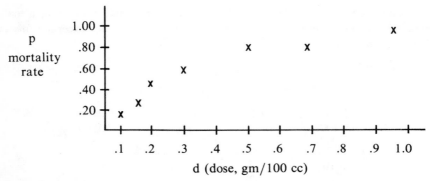

Figure 4.1. p versus d

If we suppose d was a *normal* random variable then as you know a plot of p versus d should look like

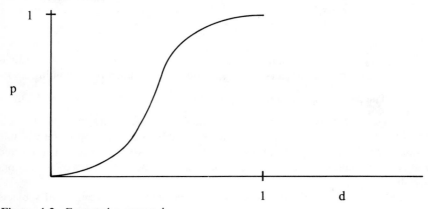

Figure 4.2. Expected p versus d

Often this is referred to as a sigmoid curve. We do not have this situation, our pattern is almost parabolic.

Usually in an attempt to obtain a graph similar to Figure 4.2 transformations on the variable d are made. Suppose for this case

$$X = \log_{10} (100 \ d). \tag{4.1}$$

Such a transformed measure of dose is often referred to as the *dose metameter*. (The 100 was used only to eliminate the case of negative numbers.) Final estimates however should be expressed in d units.

Table 4.2
DATA OF p VERSUS DOSE METAMETER

X	p
1.000	17.0
1.176	26.4
1.301	43.6
1.477	61.5
1.699	82.6
1.845	92.6
1.978	96.2

Figure 4.3 is a plot of p versus X, and is similar to Figure 4.2.

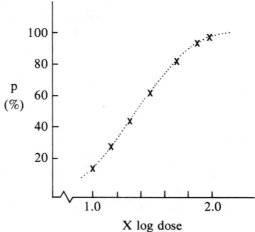

Figure 4.3.   Plot of p versus X.

We now introduce a fundamental transformation which is the basis of subsequent analyses.

4.3   *The probit transformation*

The following explanation shows how to convert such sigmoid type curve representations to straight line representations.

The *distribution* function of a *normal* random variable X can be written as

$$P = F(x) = P[X \leqslant x] = \int_{-\infty}^{x} f(t)\,dt \tag{4.2}$$

where

$$f(t) = \frac{e^{-1/2\left(\frac{t-\mu}{\sigma}\right)^2}}{\sigma\sqrt{2\pi}}, \quad -\infty < t < \infty,\; 0 < \sigma < \infty,\; |\mu| < \infty. \tag{4.3}$$

For each x, P represents the area under the normal curve below x. Conversely, given P we can obtain the corresponding value of X. For example, if P = 0.16 then, regardless of the units of X,

$$Z = \frac{X - \mu}{\sigma} \tag{4.4}$$

equals -1. Z is called the *standard normal variate*. If we let

$$Y = Z + 5, \tag{4.5}$$

then, for example, if P = 0.16 then Y = +4, or if P = 0.50 then Y = 5. Figure 4.4 shows the relationship between X and Y.

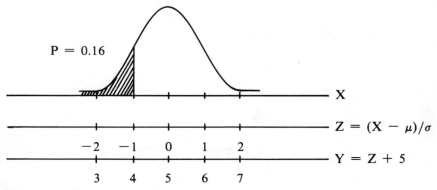

Figure 4.4. Distribution of Y.

Before we saw that p acted as an empirical cumulative relative frequency, and in this sense p can be associated with P, the true cumulative probability. Given p we can estimate P which yields a value of Y, through the transformation $Y = (X - \mu)/\sigma + 5$.

The point is that we can relate a Y value (a transformed expected X value) with the original X value through p (which is P under the normality assumption). Since

$$Y = \frac{X - \mu}{\sigma} + 5 = a + bX \tag{4.6}$$

then

$$a = 5 - \frac{\mu}{\sigma} \tag{4.7}$$

$$b = \frac{1}{\sigma}. \tag{4.8}$$

We call Y the *probit* of p; that is,

$$Y = \text{probit}(p) = a + bX, \tag{4.9}$$

where a and b are defined by (4.7) and (4.8). The word probit is a shortened form of the phrase probability unit. Of course, a probit is not a probability! Rather, it is a function of the mortality rate and this function is a linear function of X as described in (4.9). If we plot Y against X we should expect to obtain a straight line, provided the distribution of X is approximately normal. Notice that $b = 1/\sigma$ is the slope of this line and σ is the true standard deviation of X. Thus the estimate of b is a good measure of the heterogeneity of the subjects under investigation. In fact, the steeper the slope, the larger the value of b; the smaller the value of σ, the smaller the variability; and vice versa. It is important to recognize that for such a plot we are plotting Y, the expected log dose value given the observed cumulative proportion, against X, the observed log dose value which is usually controlled by the experimenter.

For a given value of p, the response rate, how do we get the value of Y, the probit of p? Well, for a given value of p one would look up in a table of normal values that value of Z that would generate such a value, and then add 5 to obtain Y. Since this is tedious for many values of p, we have provided in the appendix, Table 1. This table gives the value of $Y = \text{probit}(p)$ for $p = 0.1$ to 99.9 percent. The table was constructed from a new algorithm and computer program developed by the author. A very abbreviated form of that table is given in Table 4.3. Notice that the probit(50) = 5.

For our fruit fly example (see Table 4.1) we can now obtain the probits for the observed percentage mortality rates. This is given in Table 4.4. We have used the probit tables in the appendix and notice that we have used four significant figures in the values of X and Y.

Table 4.3
PROBIT OF p

p (%)	0	1	2	3	4	5	6	7	8	9
				INCREMENTS						
0	—	2.67	2.95	3.12	3.25	3.36	3.45	3.52	3.59	3.66
10	3.72	3.77	3.82	3.87	3.92	3.96	4.01	4.05	4.08	4.12
20	4.16	4.19	4.23	4.26	4.29	4.33	4.36	4.39	4.42	4.45
30	4.48	4.50	4.53	4.56	4.59	4.61	4.64	4.67	4.69	4.72
40	4.75	4.77	4.80	4.82	4.85	4.87	4.90	4.92	4.95	4.97
50	5.00	5.03	5.05	5.08	5.10	5.13	5.15	5.18	5.20	5.23
60	5.25	5.28	5.31	5.33	5.36	5.39	5.41	5.44	5.47	5.50
70	5.52	5.55	5.58	5.61	5.64	5.67	5.71	5.74	5.77	5.81
80	5.84	5.88	5.92	5.95	5.99	6.04	6.08	6.13	6.18	6.23
90	6.28	6.34	6.41	6.48	6.55	6.64	6.75	6.88	7.05	7.33

Table 4.4
PROBITS FOR FRUIT FLY EXAMPLE

X	p	Y
1.000	17.0	4.046
1.176	26.4	4.369
1.301	43.6	4.839
1.477	61.5	5.292
1.699	82.6	5.938
1.845	92.6	6.447
1.978	96.2	6.775

A scatter plot of Y versus X would graphically represent the relationship between Y and X. If our assumptions have been valid a straight line pattern should be evident.

4.4 *The Probit Line*

Let the model for the unknown probit line, given X, be

$$Y = a + bX + \epsilon. \qquad (4.10)$$

Using the least squares method we obtain the following estimates for a and b:

$$\hat{b} = \frac{\Sigma XY - \dfrac{1}{k} \Sigma X \, \Sigma Y}{\Sigma X^2 - \dfrac{1}{k} (\Sigma X)^2} = \frac{\Sigma xy}{\Sigma x^2}, \ \hat{a} = \frac{1}{k} (\Sigma Y - \hat{b} \, \Sigma X), \qquad (4.11)$$

where $x_i = X_i - \overline{X}$, $y_i = Y_i - \overline{Y}$, k is the sample size, the number of pairs (X_i, Y_i).

For the data of Table 4.4 we obtain: $\Sigma X = 10.476$, $\Sigma X^2 = 16.4602$, $\Sigma x^2 = 0.7821$, $\Sigma Y = 37.706$, $\Sigma Y^2 = 209.6037$, $\Sigma y^2 = 6.4977$, $\Sigma XY = 58.6801$, $\Sigma xy = 2.2504$, $\hat{b} = 2.8772$, $\hat{a} = 1.0806$, $(r = 0.998)$. Hence the estimated regression line, which we will refer to as the *probit line* is

$$Y' = 1.081 + 2.877X.$$

This probit line should now be superimposed on the scatter plot of Y on X to visualize the fit. (A test for significant regression would also be appropriate! See section 4.7 for testing the adequacy of the fit.)

4.5 *Estimation of LD50*

The probit line can now be used to obtain in a simple fashion an estimate of the quantity LD50, the median lethal dose. Recall that LD50 is that dose (or concentration) estimated to produce the effect in 50% of the population of individuals. Because of the normal structure the mean effective log dose is the median effective log dose. By definition therefore we can estimate LD50 from the fitted straight line; viz., if $Y' = \text{probit (p)} = 5$ then m, the estimate of log LD50, is

$$m = \frac{Y' - \hat{a}}{\hat{b}} = \frac{5 - 1.0806}{2.8772} = 1.3622 \tag{4.12}$$

which in original units yields:

$$LD50 = \text{antilog}(1.362) = 23.03 \text{ gm}/10^4 \text{ cc} = .23 \text{ gm}/100 \text{ cc}.$$

4.6 *Estimation of σ*

The estimate of the true slope, b, is 2.877 which in turn estimates $1/\sigma$, the standard deviation of the logarithm of the individual effective dose; here $\hat{\sigma} = 1/2.877 = 0.348$ log units or 0.022 in the original units. This quantity is a measure of the variability of the individuals, i.e., the degree to which they do not have the same individual effective dose.

4.7 *Testing the Adequacy of the Probit Line*

Table 4.5 conveniently shows the steps for testing the goodness-of-fit of the probit line. (This is an approximate procedure and it requires the values of n to be at least 30 to be statistically justified.) The null hypothesis is that the probit line adequately models the data.

The Y' column is generated from the estimate of the probit line by using the X column. The P values are obtained from the Y' column using the probit tables in an inverse manner (see Table 1 in the appendix). The sum of the values in the last column yields

$$\chi^2 = \sum \frac{(r - nP)^2}{nP(1 - P)} = 0.51. \tag{4.13}$$

Table 4.5
TEST OF FIT

X (observed)	Y' (expected)	P (expected)	n	r	nP	r − nP	$\frac{(r - nP)^2}{nP(1 - P)}$
1.000	3.958	.1487	47	8	6.97	+1.01	.172
1.176	4.464	.2958	53	14	15.68	−1.68	.255
1.301	4.824	.4300	55	24	23.65	+0.35	.009
1.477	5.330	.6295	52	32	32.73	−0.73	.004
1.699	5.969	.8338	46	38	38.35	−0.35	.020
1.845	6.389	.9175	54	50	49.96	+0.46	.051
1.978	6.772	.9627	52	50	50.06	−0.06	.002

Since the degrees of freedom (d.f.) of this χ^2 test statistic is equal to the number of values "freely" estimated (usually this means the number of independent comparisons estimated) then d.f. = $7 - 2 = 5$. (There are seven comparisons and two parameters estimated in the probit line.) Since $\chi^2_{0.05}(5) = 11.07$, then the estimated probit line is a very satisfactory model.

The reason this test statistic has a χ^2 distribution is based on the following argument: If Z_i are N(0, 1) random variables, then from distribution theory $\sum_{i=1}^{k} Z_i^2$ is a χ^2 random variable with k degrees of freedom. Now, since each r_i (number of responses) is a binomial random variable, then $Z_i = (r_i - E[r_i])/\sqrt{V[r_i]}$ is approximately a standard normal random variable, provided n_i is large. Therefore $\sum Z_i^2$ is a χ^2 variable; i.e.,

$$\chi^2 = \sum Z_i^2 = \sum \left(\frac{r_i - E[r_i]}{\sqrt{V[r_i]}} \right)^2 = \sum \left(\frac{r - nP}{\sqrt{nP(1 - P)}} \right)^2.$$

4.8 Fitting Problems

There is a real statistical problem of how to fit the best line. We of course used ordinary regression techniques and as you will recall these methods require a strong assumption: the variability of the observations (the values of Y) is the same for each value of X. For probit plots, however, it can be argued that in general this is not the situation. In fact it can be shown that the best way to fit the line is to weight each probit value Y with the weight $1/\sigma_Y^2$, where σ_Y^2 is the true variance of the probit variate Y. (See equation (5.31) for the weighted approach to the regression line.)

4.9   *Finding Confidence Limits by the Weighted Method*

As mentioned earlier the distribution of r, the number of responses out of n subjects per dosage is binomial in nature when all n members of a group respond independently to that dosage. Since the true variance of r is nPQ, then for p (the proportion of responses) the weight is $W = PQ/n$, where P is the probability of a subject responding $(Q = 1 - P)$.

The *information* or *invariance* of a variable is the reciprocal of its variance; it is used as a weighting measure in estimation theory. Therefore, the weight usually assigned to a proportion (that is, p) of a group is nw where $w = 1/PQ$.

For $Y = $ probit(p) it can be shown that the corresponding weight is nw where

$$w = f^2/PQ \qquad (4.14)$$

where f is the ordinate (frequency domain) value in a standard normal frequency function for a given value of Y, that is

$$f = \frac{1}{\sqrt{2\pi}} \, e^{-(Y-5)^2/2}$$

and P is estimated by the observed response rate p. For example, if $Y = 5$ (then p $= 50\%$ since $5 = $ probit(50)) then the weighting coefficient w is

$$w = \left(\frac{e^{-(5-5)^2/2}}{\sqrt{2\pi}} \right)^2 \frac{1}{(.5)(.5)} = \frac{2}{\pi} = 0.637.$$

The point is that for each probit value, Y, we can compute its associated weighting coefficient. A computer program was written to calculate the values of w for values of Y equal to 1.00 to 8.99 in Table 2 in the appendix. A very abbreviated form of this table is given in Table 4.6 below.

Table 4.6
w, THE WEIGHTING COEFFICIENTS FOR A PROBIT Y

Y	0.0	0.1	0.2	0.3	0.4	0.5	0.6	0.7	0.8	0.9
1	0.001	0.001	0.001	0.002	0.002	0.003	0.005	0.006	0.008	0.011
2	0.015	0.019	0.025	0.031	0.040	0.050	0.062	0.076	0.092	0.110
3	0.131	0.154	0.180	0.208	0.238	0.269	0.302	0.336	0.370	0.405
4	0.439	0.471	0.503	0.532	0.558	0.581	0.601	0.616	0.627	0.634
5	0.637	0.634	0.627	0.616	0.601	0.581	0.558	0.532	0.503	0.471
6	0.439	0.405	0.370	0.336	0.302	0.269	0.238	0.208	0.180	0.154
7	0.131	0.110	0.092	0.076	0.062	0.050	0.040	0.031	0.025	0.019
8	0.015	0.011	0.008	0.006	0.005	0.003	0.002	0.002	0.001	0.001

The standard error of the estimator m using this weighting procedure can be shown to be

$$s_m = \frac{1}{\hat{b}} \sqrt{\frac{1}{\Sigma nw} + \frac{(m - \overline{X})^2}{\Sigma nw\ (X - \overline{X})^2}},$$ (4.15)

where

$$\overline{X} = \frac{\Sigma nwX}{\Sigma nw} \quad \text{and} \quad \Sigma nw\ (X - \overline{X})^2 = \Sigma nwX^2 - \frac{(\Sigma nwX)^2}{\Sigma nw},$$ (4.16)

and m = log LD50 from probit analysis as described in section 4.5.

For example, for the data which began in Table 4.1, we can calculate s_m by constructing this convenient table:

Table 4.7
CALCULATION TABLE

X	n	Y	w	nw	nwX	nwX²
1.00	47	4.05	.4552	21.3944	21.3944	21.3944
1.18	53	4.37	.5503	29.1659	34.4158	40.6106
1.30	55	4.84	.6307	34.6885	45.0951	58.6236
1.48	52	5.29	.6174	32.0372	47.4151	70.1743
1.70	46	5.94	.4585	21.0910	35.8547	60.9530
1.85	54	6.45	.2854	15.4116	28.5115	52.7462
1.98	52	6.78	.1853	9.6356	19.0785	37.7754
Sum	359	—	—	163.4242	231.7651	342.2775

Therefore:

$$\overline{X} = \frac{231.7651}{163.4242} = 1.4182$$

$$\Sigma nw\ (X - \overline{X})^2 = 342.2775 - \frac{(231.7651)^2}{163.4242} = 13.5926$$

$$s_m = \frac{1}{2.8772} \sqrt{\frac{1}{163.4242} + \frac{(1.3622 - 1.4182)^2}{13.5926}} = 0.0277.$$

Thus we have as the estimate of log LD50:

m ± s_m = 1.3622 ± 0.0277 or 1.36 ± 0.03.

The approximate 95% confidence limits are written as

m ± 1.96 s_m = 1.3622 ± 0.0543, i.e., (1.308, 1.417),

and so for LD50 we obtain (20.32, 26.09), which in the original units becomes (0.203, 0.261) gm/100 cc.

4.10   *Abbott's Formula*

A zero dose level defines a control group. Since the logarithm of zero is minus infinity, then X, the log dose value, cannot be used in the ordinary regression analysis in probit analysis. When there are no mortalities for this group, there is no information, and therefore this dose value is not included in the analysis. However, when there are mortalities, this information can be used to adjust the observed mortality rates for the other dose levels.

Suppose C is the "natural mortality rate" of the subjects under study and for the moment, let p_i' denote the i-th mortality rate, for the non-zero dose levels which are actually observed, and let p_i be the i-th true mortality rate for those subjects who die due to the i-th level of non-zero dose. Then it follows that

$$p_i' = C + (1-C)p_i.$$

Therefore,

$$p_i = \frac{p_i' - C}{1 - C}.$$ (4.17)

This formula is known as Abbott's formula and is used to adjust the observed mortality rates p_i' by an estimate of the natural mortality rate C from a control group. (See Booth, 1975.) The new values p_i are then used in the probit analysis. The formula (4.17) is only applied to those $p_i' \geq C$.

The following data has been slightly altered from Martin (1940); also, see Finney (1971) and Ashton (1972). This toxicity experiment concerns the derris root as it affects the grain beetle (Oryzaephilus surinamensis). The dosage is in mg of dry root per litre, and r is the number of mortalities out of n beetles.

d	n	r	p'
0	129	21	0.163
10	126	58	0.460
50	128	115	0.898
100	127	124	0.976
150	142	141	0.993

The first dose level acts as a control group and the value 0.163 estimates C, the natural mortality. We now apply Abbott's formula (4.17) to obtain the following data:

X	d	p	Y
1.000	10	0.355	4.6286
1.699	50	0.879	6.1701
2.000	100	0.972	6.9115
2.176	150	0.992	7.4093

Simple regression analysis yields the probit line $Y' = 2.26 + 2.34X$. Then LC50 = 14.8 mg per litre.

Abbott's formula is only applied to observed mortality rates which exceed the estimate of the natural mortality rate.

4.11   *Berkson's Adjustment*

An observed mortality rate of 0 or 100 percent yields a probit value unsuitable for regression analysis. Berkson has suggested that if a non-zero dose level has a zero mortality rate for n subjects, then this zero value should be replaced by $1/2n$; and if a 100% rate is observed for n subjects this value should be replaced by $1 - (1/2n)$.

4.12   *Remark*

We have established how to find both the point estimate and the interval estimate of the median lethal dose for this class of quantal assays. The next chapter introduces other methods.

After chapter 5 there are many exercises; they are directly related to probit analysis, and almost all have been taken from the scientific literature. They cover a very wide range of experiments in the biological sciences and it is for this reason that the reader is strongly advised to read these exercises to experience the type of problems associated with the techniques described in this section as well as those described in chapter 5.

5 Other Methods for Estimating LC50

Here is a list of other methods that have appeared in the literature for estimating the parameter LC50:

1. The Dragstedt-Behrens method
2. The Spearman-Kärber method
3. The extreme lethal dosage method
4. The Reed-Muench method
5. The Thompson moving average method
6. The maximum likelihood method
7. The Litchfield-Wilcoxon graphical method
8. The Dixon-Mood staircase method
9. The Shuster-Dietrich method
10. The Shuster-Yang method.

There are still other methods which we could enumerate, but they require more statistical background than is required for this book. However, for those who are very statistically inclined, see Kraft and van Eeden (1964), Ramsey (1972), Chmiel (1976), Wesley (1976), Basu and Fagerstrom (1979) and Bhattachary (1981) for a Bayesian approach; Cochran and Davis (1963), (1964), Wetherill (1963), Wetherill et al (1966), Davis (1971), Freeman (1970) and Tsutakawa (1972), (1980) for a sequential approach; and van Eeden (1960) and Davis (1972) for a discussion of distribution-free methods.

We now will describe in brief terms each of these 10 methods.

5.1 *The Dragstedt-Behrens method*

Dragstedt and Lang (1928) and independently by Behrens (1929) suggested the following method. At each dose X let T_1 be total number of subjects responding at doses $\leq$ X and let T_2 be total number of subjects not responding at doses $\geq$ X. Use $T_1/(T_1 + T_2)$ as the new estimates of response rate at dose X. With this set of new values use graphical methods to get LC50. Finney (1971) states that this "method may behave reasonably well if the data are equally spaced in X, have equal n at each X, and are moderately symmetrical" and that its "only merit" is its numerical simplicity.

Here is an artificial example to illustrate the calculations for the Dragstedt-Behrens method. Suppose the toxicity of some poison is being studied with rats as the subjects, and ten rats were exposed to each of eight different dosages. In the last column p* denotes the estimate of the cumulated percentage mortality rate; i.e., $p^* = T_1/(T_1 + T_2)$ or using the notation of the table $p^* = (\Sigma r)/T$.

Table 5.1

dose d	$\log_{10}$ dose X	n	no. dead r	no. alive n − r	Σr	$\Sigma(n-r)$	Total T	$p^* \cdot 100$
1.0	.0000	10	0	10	0	44	44	.00
1.5	.1761	10	1	9	1	34	35	2.86
2.0	.3010	10	2	8	3	25	28	10.71
2.5	.3979	10	4	6	7	17	24	29.17
3.0	.4771	10	6	4	13	11	24	54.17
3.5	.5441	10	7	3	20	7	27	74.07
4.0	.6021	10	7	3	27	4	31	87.10
4.5	.6532	10	9	1	36	1	37	97.30
Totals		80	36	44	—	—	—	—

By linear interpolation we can find that value of X corresponding to $p^* = 50\%$. It follows that an approximate estimate of log LC50 is m = 0.4639; therefore LC50 = 2.91 units.

A crude estimate of the standard error can be shown to be

$$SE(m) = \sqrt{\frac{0.79 \ (h) \ IR}{n}}$$

where h = (maximum log dose − minimum log dose)/k; k = number of intervals; IR is the interquartile range = log LC75 − log LC25; and n is the number of subjects per dose (if this value varies from dose to dose we suggest n = integer part of the average value).

For the data of Table 5.1, we obtain the estimates: log LC75 = 0.5482, log LC25 = 0.3670, so that

$$SE(m) = \sqrt{\frac{0.79(0.0933)(0.1722)}{10}} = 0.0356.$$

The very approximate 95% confidence interval can be taken as

$$m \pm 1.96 \ SE(m) = 0.4639 \pm 1.96(0.0356) = 0.394 \text{ to } 0.534$$

which in the original dose units becomes (2.48, 3.42).

For the data considered in section 4 (see Table 4.1) we have:

Table 5.2

log dose X	no. of subjects n	no. dead r	no. alive n − r	Σr (down)	Σ(n−r) (up)	Total T	Acc. % Mortality $\frac{\Sigma r}{T} \times 100$
1.00	47	8	39	╱ 8	143	151	5.30
1.18	53	14	39	22	104	126	17.46
1.30	55	24	31	46	65	111	41.44
1.48	52	32	20	78	34	112	69.64
1.70	46	38	8	116	14	130	89.23
1.85	54	50	4	166	6	172	96.51
1.98	52	50	2	216	2	218	99.08
Totals	359	216	143	—	—	—	—

Since we want to obtain that value of X corresponding to a p* = 50%, then

$$m = 1.3 + 0.18 \left(\frac{50 - 41.44}{69.44 - 41.44} \right) = 1.3546$$

or $\widehat{LC50} = 22.63$ gm/10^4 cc = 0.23 gm/100 cc.

To obtain SE(m) we calculate:

h = (1.98 − 1.00)/7 = 0.14

$\widehat{LC75} = 1.5402$

$\widehat{LC25} = 1.2492$

IR = 0.2910

n = [359/7] = 51.

Therefore

$$SE(m) = \sqrt{(.79)(.14)(.291)/51} = 0.02512$$

The approximate 95% confidence interval becomes m ± 1.96 SE(m) = 1.355 ± 0.049 or (1.306, 1.404). Thus for LC50 we have (0.202, 0.253) gm/100 cc. (Notice that this is quite close to the interval obtained by the weighted method in section 4.9.)

5.2   The Spearman-Kärber Method

Kärber (1931) independently of Spearman (1908) suggested an estimator of log LC50 is

$$m = \frac{1}{2} \sum_{i=1}^{k-1} (p_{i+1} - p_i)(X_i + X_{i+1}), \tag{5.1}$$

where X_i are the log dose for which r_i out of n_i subjects respond and $p_i = r_i/n_i$ with the condition that $p_1 = 0\%$ and $p_k = 100\%$. If the doses are equally spaced so that $X_{i+1} - X_i = d$, then (5.1) reduces to the following simple expression (see Appendix 8 for proof):

$$m = X_k + d \left(\frac{1}{2} - \sum_{i=1}^{k} p_i \right).$$ (5.2)

If the number of subjects per dose is also constant then we use

$$m = X_k + d \left(\frac{1}{2} - \frac{1}{n} \sum_{i=1}^{k} r_i \right).$$ (5.3)

In this latter case, the best form of the standard error is

$$SE(m) = \frac{d}{n} \sqrt{\frac{\Sigma r_i (n - r_i)}{n - 1}}.$$ (5.4)

Consider the data of Harris (1959) who studied the toxicity of the insecticide dieldrin on fathead minnows (*P. promelas*). The exposure time was 72 hours and 10 fish were used per dose.

Table 5.3

X, log dose	r
.125	0
.250	3
.375	5
.500	8
.625	10
.750	10

Then $m = \log LC50 = 0.750 + 0.125 \left(\frac{1}{2} - \frac{36}{10} \right) = 0.363$

with $SE(m) \simeq 0.0328$.

The derivation of these formulae requires that the doses extend over the whole range from zero to 100% response and that the increase in response rate between two successive doses is concentrated at the centre of the interval. Equal spacing of doses or equal number of subjects per dose is not required, and the calculations are not complicated. Recent developments in this method appear in Chang and Johnson (1972), Church and Cobb (1971), (1973), (1976), Finney (1964), Pursey (1977), Chmiel (1976), Hamilton (1979), (1980), Hamilton et al (1977) and Miller and Halpern (1980).

5.3   *The Extreme Lethal Dosage Method of Gaddum (1933)*

This method is used only for those assays in which one subject is tested at each of a series of doses and the interval between successive log doses is constant and finally a sufficiently large number of levels of the doses is taken. The average of lowest log dose achieving a response and the highest log dose achieving a non-response is the estimate of LC50. For a deeper discussion, see Finney (1950), (1964).

5.4   *The Reed-Muench Method*

This method was suggested by Reed and Muench (1938) and because of its numerical simplicity it became quite popular. Today, however, the Reed-Muench (as well as the Dragstedt-Behrens and the extreme lethal dosage method) are not recommended procedures. In fact Finney (1964) states "(these) methods ought never be used", (page 551).

The discussion given here is for those who have seen this method and are curious on its mechanics. From the analysis it will be evident that this procedure assumes that any subject responding to a given dose of an agent would respond to all higher doses; and that any subject not responding to a given dose would not respond to a lower dose.

If $T_1(r, X)$ is the total of all values of r for log doses equal to or less than X and $T_2(n - r, X)$ is the total of all values $(n - r)$ for log doses equal to or greater than X, then if one of the doses used in the experiment has $T_1(r_i, X_i) = T_2(n_i - r_i, X_i)$ then X_i is taken as m, the estimate of log LC50. If the doses are equally spaced so that $X_{i+1} - X_i = d$ for all i, if the number of subjects per dose is constant, and if the interpolation is to be made between doses X_i and X_{i+1} then the estimate of log LC50, m, is the solution of

$$T_1 + r_{i+1}(m - X_i)/d = T_2 - (n - r_i)(m - X_i)/d \tag{5.5}$$

or

$$m = X_i + \frac{d(T_2 - T_1)}{n - r_i + r_{i+1}} \tag{5.6}$$

where $T_2 = T_2(n - r_i, X_i)$ and $T_1 = T_1(r_i, X_i)$. Consider the example considered earlier in Table 5.3. For this data the totals are given in the table below.

Table 5.4

X	n	r	T_1	n − r	T_2
.125	10	0	0	10	24
.250	10	3	3	7	14
.375	10	5	8	5	7
.500	10	8	16	2	2
.625	10	10	26	0	0
.750	10	10	36	0	0

Then:

$$m = 0.250 + \frac{0.125(14 - 3)}{10 - 3 + 5} = 0.3646.$$

Notice that the method is similar to a double linear interpolation. Here $T_1 < T_2$ at X_2 and $T_1 > T_2$ at X_3. A pictorial representation of the estimation technique is:

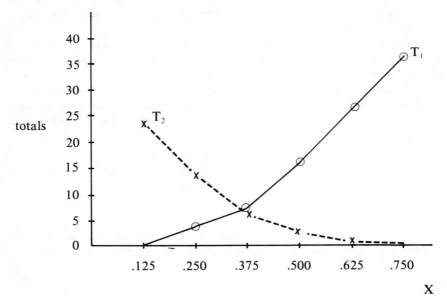

Here is another example which was taken from Finney (1964), p. 535, where n = 6:

Table 5.5

X	r	T_1	n − 4	T_2
1	0	0	6	32
2	0	0	6	26
3	1	1	5	20
4	0	1	6	15
5	2	3	4	9
6	4	7	2	5
7	4	11	2	3
8	6	17	0	1
9	5	22	1	1

Then

$$m = 5 + \frac{1(9 - 3)}{6 - 2 + 4} = 5.75.$$

5.5 *The Thompson Moving Average Method*

Thompson's (1947) method utilizes moving averages. For example, if for each i (i = 1, 2, . . ., k) we plot $(p_i + p_{i+1})/2$ versus $(X_i + X_{i+1})/2$ and then by simple linear interpolation find that value of X corresponding to 50% response, we have log LC50.

The merits of these first five methods have been discussed in great detail by Finney (1950), (1953), (1964), (1971) and Armitage and Allen (1950).

5.6 *The Litchfield-Wilcoxon (1949) Method*

This is a rapid crude graphical approximation and uses a line drawn by eye to fit points plotted on logarithmic probability paper for each dose and response. This method does not take into account the weighting component of the points, and it requires the points to be near the line and neglects those responses near 0 or 100%. For "good" data Finney (1971) says "the results are often very close to those from maximum likelihood estimation". Other such approximation methods can be found in Abbott (1925), Weil (1952), Burdick (1957), Cochran and Davis (1964), Shorack (1966) and Cornell and Peterson (1970). Recent biological papers utilizing this method are Brett (1952), Sprague (1969), Sanders (1970), Gill et al. (1970), McLeese (1974a), (1974b), Vincent and Lindgren (1975), and Davis and Hoos (1975). See exercise 5.9 and 5.12.

5.7   *The Dixon-Mood (1948) Method*

This method is also known as the "staircase method" or "up and down method". The experiment begins by testing one subject at log dose X_0 (believed to be about the value of log LC50). Then a series of equally spaced log doses are chosen, say . . ., X_{-2}, X_{-1}, X_0, X_{+1}, X_{+2}, . . . and the following procedure is followed: if the subject dies the next subject is tested at a dose one step lower, and if it lives the next subject is tested at a dose one step higher and so on. This procedure is "sequential" and assumes that later doses tend to concentrate about the true value of LC50. Their estimator of the log LC50 is

$$m = X_0 + d\left(\frac{\Sigma \, i \cdot r_i}{n_r} - \frac{1}{2} \right), \tag{5.7}$$

where i is the subscript of X_i, r_i is the number of responses to log dose X_i and n_r is the total number of responses.

For example, suppose $+$ = response and 0 = non-response and the results were

i	X_i	Response types							Totals +	Totals 0
1	1.5		+		+				2	0
0	1.1	+	+	0	+	0	+	0	4	3
−1	0.7	0	0		0		+	+	2	3
−2	0.3						0		0	1
									8	7

Therefore

$$m = 1.1 + 0.4 \left(\frac{1(2) + 0(4) - 1(2) - 2(0)}{8} - \frac{1}{2} \right) = 0.9.$$

At this time no more will be said about this method. Probably the best references on this interesting approach are Brownlee et al. (1953), Robbins and Monro (1951), Cochran and Davis (1965), Dixon (1965), Davis (1971), Hsi (1969) and recently Little (1974a), (1974b), (1975).

5.8   *The Shuster-Dietrich Method*

Recently Shuster and Dietrich (1976) have proposed a general inverse regression procedure for estimating dose-response curves in quantal response assays. For the *linear* case they have proven the following theorem, using the notation of this chapter:

Theorem 5.1

The estimate of $\log_{10}$ LC50, m, is given by

$$m = \tilde{Y} - \hat{b}(\tilde{Z} - 45) \tag{5.8}$$

where

$i \quad = 1, 2, \ldots, k$

$X_i \quad = \log_{10} d_i$

$Y_i \quad = (n_i/n)^{1/2} X_i$

$$n \quad = \sum_{i=1}^{k} n_i$$

$$\tilde{Y} \quad = \sum_{i=1}^{k} (n_i/n)^{1/2} Y_i = \frac{1}{n} \sum_{i=1}^{k} n_i \log d_i$$

$Z_i \quad = (n_i/n)^{1/2} \sin^{-1} (\hat{p}_i^{1/2})$
(where the units of the inverse sine function is in degrees rather than radians)

$$\tilde{Z} \quad = \sum_{i=1}^{k} (n_i/n)^{1/2} Z_i = \frac{1}{n} \sum_{i=1}^{k} n_i \sin^{-1} (\hat{p}_i^{1/2})$$

$$\hat{b} = S_{YZ}/S_{ZZ}$$

$$S_{YZ} = \sum_{i=1}^{k} Y_i Z_i - \tilde{Y} \tilde{Z}, \ S_{ZZ} = \sum_{i=1}^{k} Z_i^2 - (\tilde{Z})^2.$$

The response-dose model which they start with is of the form

$$X_i = a + b(\sin^{-1} (p_i^{1/2})) \tag{5.9}$$

where $X_i = \log_{10} d_i$ (the log dose metameter) and p_i is the true fraction of responses (not percentages) for which $\hat{p}_i$ is the observed value.

Let us illustrate this method using the data of the example appearing in sections 4.2 and 5.1. Recall that in section 4.2, probit analysis suggested that $L\hat{C}50 = 0.230$ gm/100 cc. and that in section 5.1, the Dragstedt-Behrens method suggested $L\hat{C}50 = 0.226$ gm/100 cc. In the calculations below, the Shuster-Dietrich inverse linear regression method suggests $L\hat{C}50 = 0.236$ gm/100 cc.

Table 5.6

DATA AND CALCULATIONS

d_i	n_i	r_i	$\hat{p}_i$	X_i	Y_i	$n_i X_i$	$\sin^{-1}(\hat{p}_i^{1/2})$	Z_i
.10	47	8	.170	1.00	.3618	47.00	24.350	9.253
.15	53	14	.264	1.18	.4519	62.33	30.918	11.880
.20	55	24	.436	1.30	.5092	71.56	41.323	16.174
.30	52	32	.615	1.48	.5622	76.81	51.649	19.657
.50	46	38	.826	1.70	.6082	78.15	65.346	23.391
.70	54	50	.926	1.85	.7156	99.65	74.215	28.783
.95	52	50	.962	1.98	.7527	102.84	78.759	29.974
	359					538.34		

Note:

$k = 7$, that is, $i = 1, 2, \ldots, k$

$X_i = \log_{10}(100\, d_i)$

$n \quad = \Sigma\, n_i = 359$

$Y_i = (n_i/n)^{1/2}\, X_i$

$\tilde{Y} \quad = \Sigma\, \sqrt{(n_i/n)}\, Y_i = \dfrac{1}{n}\Sigma\, n_i\, X_i = \dfrac{1}{359}(538.33) = 1.4995$

$Z_i = (n_i/n)^{1/2} \sin^{-1}(\hat{p}_i^{1/2})$

$\tilde{Z} \quad = \Sigma\,(n_i/n)^{1/2}\, Z_i = \dfrac{1}{n}\Sigma\, n_i \sin^{-1}(\hat{p}_i^{1/2}) = \dfrac{18,850.593}{359} = 52.509$

$S_{YZ} = \Sigma\, Y_i\, Z_i - \tilde{Y}\,\tilde{Z} = 85.38812 - 78.73770 = 6.65042$

$S_{ZZ} = \Sigma\, Z_i^2 - (\tilde{Z})^2 = 3148.84436 - (52.5086)^2 = 391.69128$

$\hat{b} \quad = \dfrac{S_{YZ}}{S_{ZZ}} = \dfrac{6.6504}{391.6913} = 0.01698.$

So the estimate of the log median lethal dosage is

$$m = \tilde{Y} - \hat{b}(\tilde{Z} - 45) = 1.450 - 0.01698\,(52.50 - 45) = 1.372;$$

and so $\hat{LC50} = 10^{1.372} = 23.55$ gm/10^4 cc., or, $\hat{LC50} = 0.236$ gm/100 cc.

It can be shown that the approximate 95% confidence limits for the log LC50 are given by

$$m \pm 1.96\, \frac{S_n}{\sqrt{n}} \qquad\qquad (5.10)$$

where

$$S_n^2 = 820.7 \left\{ (\tilde{Z} - 45)^2 \cdot \frac{S_{YY}}{S_{\tilde{Z}Z}^2} + \hat{b}^2 \right\}. \tag{5.11}$$

For our example:

$$S_{YY} = \Sigma \, Y_i^2 - (\tilde{Y})^2 = 2.35899 - (1.4995)^2 = 0.1104298$$

$$S_n^2 = 820.7 \left\{ (52.509 - 45)^2 \cdot \frac{0.11043}{(391.69128)^2} + (0.01698)^2 \right\} = 0.26992.$$

$$1.96 \, \frac{S_n}{\sqrt{n}} = 1.96 \, \frac{0.51953}{\sqrt{359}} = 0.0537$$

$$m - 1.96 \, \frac{S_n}{\sqrt{n}} = 1.318$$

$$m + 1.96 \, \frac{S_n}{\sqrt{n}} = 1.426.$$

Hence the approximate 95% confidence limits for LC50 is (20.78, 26.65) gm/10^4 cc. or (.208, .267) gm/100 cc. These are consistent and comparable with (.203, .260) for the weighted probit method, and with (.202, .253) for the Dragstedt-Behrens method.

5.9   *The Shuster-Yang (1975) Method*

This method provides a simple algorithm for locating the interval in which LC50 is likely to lie. From this interval the estimate of LC50 is found by linear interpolation.

The method requires that the response rate increases as the dosage increases and that the subjects react independently. Since interpolation is used, this procedure does not construct a distribution-free estimator for LC50 but rather makes a distribution-free statement about the interval in which LC50 should lie.

Suppose there are k dose levels with corresponding response rates p_1, p_2, . . ., p_k. Without loss of generality we can let $p_0 = 0$ and $p_{k+1} = 1$. Define

$$A_i = r_i - n_i(.50) \text{ with } A_0 = 0, \, A_{k+1} = 0$$

$$S_i = \sum_{u=0}^{i} A_u$$

$$j = \min \{i : p_i \geq .50\}.$$

The algorithm for estimating j is

Step 1: find the A_i (i = 1, 2, . . ., k)
Step 2: find the S_i
Step 3: find the first integer s, say, such that

$$S_{s-1} = \min \{S_1, S_2, \ldots, S_k\}.$$

Then the estimate of $j, \hat{j}$, say, is defined by this s; so LC50 is likely to be in the interval (d_{s-1}, d_s).

Here is a simple example:

Table 5.7

i	d	n	r	A	S
1	10	47	8	−15.5	−15.5
2	15	53	14	−12.5	−28.0
3	20	55	24	−3.5	−31.5
4	30	52	32	6.0	−25.5
5	50	46	38	15.0	−10.5
6	70	54	50	23.0	12.5
7	95	52	50	24.0	36.5

Since the minimum S value is $S_3 = -31.5$, then $\hat{j} = 4$. Therefore LC50 is likely to be in the interval (20, 30). By linear interpolation on d and p we obtain the estimate of LC50 as

$$\text{L}\hat{\text{C}}50 = 20 + (30 - 20)\frac{0.5 - (24/55)}{(32/52) - (24/55)} \simeq 23.58$$

5.10   *The Maximum Likelihood Method*

Suppose we want to relate the percentage response with the log dose by letting $P(X = x)$ be the probability of response at log dose $X = x$. Remember that each individual in a quantal type of bioassay has a threshold value below which it will not respond to the drug and above which it will respond. For a class of individuals the distribution of these threshold values is known to be of the form

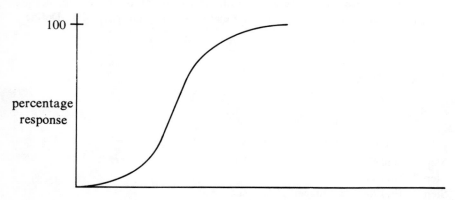

The distribution function corresponding to this curve is called the *tolerance* distribution (for a set of subjects to a particular agent).

Let $F(\alpha + \beta X)$ be this tolerance distribution function where α and β are parameters. Then the probability of a response at dose X can be written as

$$P(X) = F(\alpha + \beta X) = F\left(\frac{X - \mu}{\sigma}\right) = \int_{-\infty}^{\frac{X - \mu}{\sigma}} f(t)\, dt \qquad (5.12)$$

where $\alpha = -\dfrac{\mu}{\sigma}$ and $\beta = \dfrac{1}{\sigma}$ and $f(t)$ is the probability density function of the tolerance distribution function.

Clearly if we want to find a model for $P(X)$ it suffices to find a model for the $f(t)$. It must be stressed here that we are in essence assuming the existence of a tolerance distribution of threshold values with one or two unknown parameters.

Two candidates for $f(t)$ are well documented in the literature:

(1) *normal*

$$f(t) = \frac{1}{\sqrt{2\pi}}\, e^{-t^2/2}, \; |t| < \infty \qquad (5.13)$$

(2) *logistic* (see Berkson (1944), (1950).)

$$f(t) = \frac{e^t}{(1 + e^t)^2}, \; |t| < \infty. \qquad (5.14)$$

For the *logistic* model,

$$P(X) = F(\alpha + \beta X) = \int_{-\infty}^{\alpha + \beta X} \frac{e^t}{(1 + e^t)^2}\, dt = \frac{e^{\alpha + \beta X}}{1 + e^{\alpha + \beta X}}$$

and therefore if you want to find the value of log LC50 you would want that value of X, say X_0 for which $P(X_0) = .50$. But

$$\frac{1}{2} = \frac{e^{\alpha + \beta X_0}}{1 + e^{\alpha + \beta X_0}} \text{ implies } X_0 = -\alpha/\beta.$$

For the *normal* model

$$P(X) = F(\alpha + \beta X) = \int_{-\infty}^{\alpha + \beta X} \frac{e^{-t^2/2}}{\sqrt{2\pi}}\, dt$$

and here the corresponding value X_0 is that value satisfying

$$P\left(\frac{X_0 - \mu}{\sigma}\right) = \frac{1}{2}$$

that is $X_0 = \mu$; or equivalently, $\alpha + \beta X_0 = 0$, i.e., $X_0 = -\alpha/\beta$.

In either case we see that the log LC50 can be estimated once we obtain the estimates of α and β.

The *maximum likelihood method* argues as follows: since r is a binomial random variable, then for the ith dose level, the likelihood function is proportional to

$$P_i^{r_i} Q_i^{n_i - r_i}, \; i = 1, 2, \ldots, k, \tag{5.15}$$

so that the *likelihood function* for a quantal assay is

$$\mathcal{L}(\alpha, \beta | \text{data}) \propto \prod_{i=1}^{k} P_i^{r_i} Q_i^{n_i - r_i} \tag{5.16}$$

and the *log-likelihood function* is proportional to

$$L = \log \mathcal{L} = \sum_{i=1}^{k} (r_i \log P_i + (n_i - r_i) \log Q_i) \tag{5.17}$$

where

$$P_i = P\left[\frac{X_i - \mu}{\sigma}\right] = P(\alpha + \beta X_i)$$

$$Q_i = 1 - P\left[\frac{X_i - \mu}{\sigma}\right] = 1 - P(\alpha + \beta X_i).$$

Then the maximum likelihood (ML) estimates of the parameters α and β are the solutions of the equations

$$\frac{\partial L}{\partial \alpha} = 0 \quad \text{and} \quad \frac{\partial L}{\partial \beta} = 0. \tag{5.18}$$

Caution: in the remainder of this section we have written $\sum$ for $\sum_{i=1}^{k}$.

Now

$$\frac{\partial L}{\partial \alpha} = \sum \left(\frac{r_i}{P_i} \frac{\partial P_i}{\partial \alpha} + \frac{(n_i - r_i)}{Q_i} \frac{\partial Q_i}{\partial \alpha} \right) \tag{5.19}$$

$$\frac{\partial L}{\partial \beta} = \sum \left(\frac{r_i}{P_i} \frac{\partial P_i}{\partial \beta} + \frac{(n_i - r_i)}{Q_i} \frac{\partial Q_i}{\partial \beta} \right). \tag{5.20}$$

But in general if $P_i = P[\alpha + \beta X_i] = \displaystyle\int_{-\infty}^{\alpha+\beta X_i} f_i(t)\ dt$, $Y_i = \alpha + \beta X_i$, and $f_i = f(Y_i)$, then

$$\frac{\partial P_i}{\partial \alpha} = \frac{\partial P_i}{\partial Y_i}\ \frac{\partial Y_i}{\partial \alpha} = f_i \cdot 1$$

$$\frac{\partial P_i}{\partial \beta} = \frac{\partial P_i}{\partial Y_i} \cdot \frac{\partial Y_i}{\partial \beta} = f_i \cdot X_i$$

$$\frac{\partial Q_i}{\partial \alpha} = \frac{\partial (1 - P_i)}{\partial \alpha} = -\frac{\partial P_i}{\partial \alpha}$$

$$\frac{\partial Q_i}{\partial \beta} = -\frac{\partial P_i}{\partial \beta}.$$

Therefore:

$$\frac{\partial L}{\partial \alpha} = \sum \left[\frac{r_i}{P_i} f_i + \frac{(n_i - r_i)}{Q_i} (-f_i) \right]$$

$$\frac{\partial L}{\partial \beta} = \sum \left[\frac{r_i}{P_i} f_i X_i + \frac{(n_i - r_i)}{Q_i} (-f_i X_i) \right]$$

or,

$$\frac{\partial L}{\partial \alpha} = \sum \frac{f_i}{P_i Q_i} [r_i Q_i - (n_i - r_i) P_i]$$

$$\frac{\partial L}{\partial \beta} = \sum \frac{f_i X_i}{P_i Q_i} [r_i Q_i - (n_i - r_i) P_i].$$

Therefore if we let

$$w_i = f_i^2 / P_i Q_i \text{ and } p_i = r_i / n_i \tag{5.21}$$

then

$$\frac{\partial L}{\partial \alpha} = \sum n_i w_i ((p_i - P_i)/f_i) \tag{5.22}$$

$$\frac{\partial L}{\partial \beta} = \sum n_i w_i X_i ((p_i - P_i)/f_i). \tag{5.23}$$

Finally, equating these two equations to zero one can, at least theoretically, obtain the ML estimates of α and β, provided of course one knew the form of P_i (and thus Q_i and f_i), since X_i, p_i and n_i are known.

5.10.1 *Logistic Distribution*

If the tolerance distribution is *logistic,* then for i = 1, 2, . . ., k,

$$P_i = P(X_i) = F(\alpha + \beta X_i) = \frac{e^{\alpha + \beta X_i}}{1 + e^{\alpha + \beta X_i}}$$

$$Q_i = 1 - P_i = \frac{1}{1 + e^{\alpha + \beta X_i}}.$$

Let

$$Y_i = \alpha + \beta X_i.$$

Then

$$f_i = \frac{d}{dY_i} F(Y_i) = f(Y_i) = \frac{e^{Y_i}}{(1 + e^{Y_i})^2}.$$

Also

$$w_i = \frac{f_i^2}{P_i Q_i} = f_i.$$

Thus (5.22) reduces to

$$\sum n_i (p_i - P_i) = 0$$

or,

$$\sum r_i - \sum n_i \frac{e^{Y_i}}{1 + e^{Y_i}} = 0,$$

(where $\sum r_i$ = total number of responses) or

$$\sum r_i = \sum \frac{n_i e^{\alpha + \beta X_i}}{1 + e^{\alpha + \beta X_i}}.$$

Since

$$P_i = \frac{e^{\alpha + \beta X_i}}{1 + e^{\alpha + \beta X_i}}, \qquad (5.24)$$

then $\dfrac{\partial L}{\partial \alpha} = 0$ implies

$$\sum n_i P_i = \sum r_i. \qquad (5.25)$$

Through an analogous argument $\dfrac{\partial L}{\partial \beta} = 0$ implies

$$\sum n_i P_i X_i = \sum r_i X_i, \qquad (5.26)$$

where P_i is of course a function of α and β.

Notice that these two conditions can be written in terms of only P_i. By subtraction

$$\sum n_i \, P_i(1 - X_i) = \sum r_i(1 - X_i). \tag{5.27}$$

Thus the data yields the X_i, n_i, r_i values and so one can, at least theoretically, solve for P_i values which are in turn functions of $\hat{\alpha}$, $\hat{\beta}$, the ML estimates. An *iterative* procedure is clearly warranted (some kind of initial guess of the estimates are of course required).

5.10.2 *Normal*

Now suppose the form of the tolerance distribution is normal. Then for each dose level X_i, $i = 1, 2, \ldots, k$

$$P_i = \int_{-\infty}^{Z_i} \frac{e^{-t^2/2}}{\sqrt{2\pi}} dt, \text{ where } Z_i = \frac{X_i - \mu}{\sigma} \tag{5.28}$$

and $Y = Z + 5$, the probit. As before, (5.28) defines the relation between the stimulus X and the probability that a randomly selected subject will respond to it. As explained in section 4.2 we have the problem to estimate μ and σ where the *observed* (or "empirical") probits have different weights, $n_i \, w_i$, where n_i are the number of individuals in each dose group, and the w_i are weighting coefficients defined in section 4.9. Recall

$$w_i = \frac{f_i^2}{P_i(1 - P_i)}, \, f_i = \frac{e^{-1/2 \, Z_i^2}}{\sqrt{2\pi}}. \tag{5.29}$$

We obtain the estimators of μ and σ by fitting the regression line

$$Y = \frac{X - \mu}{\sigma} + 5 = \left(5 - \frac{\mu}{\sigma} \right) + \left(\frac{1}{\sigma} \right) X = a + bX \tag{5.30}$$

where the estimators of b and a are the *weighted* least squares estimators

$$\hat{b} = \frac{\Sigma \, nw \, (X - \overline{X})(Y - \overline{Y})}{\Sigma \, nw \, (X - \overline{X})^2}, \, \hat{a} = \overline{Y} - b\overline{X}, \text{ respectively}, \tag{5.31}$$

where

$$\overline{X} = \frac{\Sigma \, nwX}{\Sigma \, nw}, \, \overline{Y} = \frac{\Sigma \, nwY}{\Sigma \, nw}.$$

The general method for obtaining this *provisional* regression line and the estimate of log LC50 based on this maximum likelihood argument is:

(1) From the observed p_i find the ("observed") probits Y_i.
(2) Plot Y_i versus X_i and fit a straight line to the points by eye.
(3) Using this line obtain the set of "expected" probits Y_i', for each X_i.

(4) Obtain the set of "working" probits y_i'. (See Remark 1 below.)

(5) For the y_i' find the corresponding weighting coefficients w_i (or better still, the values $n_i w_i$ since they always occur together); and with the X_i find a *second* approximation to the regression line (5.30).

(6) Use this line to get a second set of expected probits Y_i'', for each X_i; and we repeat the cycle (see step (3)). This iteration is continued until there is no further change in the line. (See Remark 2 below.)

(7) From the final line, the estimate of log LC50, m, is obtained by letting $Y = 5$ and solving for X, i.e.,

$$m = \frac{Y - a}{\hat{b}} = \frac{5 - (\overline{Y} - \hat{b}\overline{X})}{\hat{b}} = \frac{5 - \overline{Y}}{\hat{b}} + \overline{X}. \qquad (5.32)$$

The SE of this estimate is given by (4.15).

Remark 1: The working probits y_i' are defined by the relation

$$y_i' = Y_i' + \frac{p_i - P_i}{f_i} \quad \text{(Notice that } y_i' \text{ is a function of } Y_i' \text{ and } p_i.) \qquad (5.33)$$

where p_i = proportion (not percentage) of responses observed

$$f_i = \frac{1}{\sqrt{2\pi}} e^{-(Y_i' - 5)^2/2} \ , \qquad\qquad P_i = \int_{-\infty}^{Y_i' - 5} \frac{e^{-t^2/2}}{\sqrt{2\pi}} \, dt$$

and Y_i' are the expected probits.

For example, if $Y' = 6.2$, $p = 0.723$ then $f = \dfrac{1}{\sqrt{2\pi}} e^{-1/2(6.2 - 5)^2} = 0.194$

$$P = \int_{-\infty}^{6.2 - 5} \frac{e^{-t^2/2}}{\sqrt{2\pi}} \, dt = 0.885 \text{ and so } y' \simeq 5.37.$$

Clearly this seems like a lot of work and if one had a computer (and a good program!) one could obtain each y' for given values of Y' and p. However, as you may have guessed by now, a computer program was developed to calculate these working probits for selected values of Y and p: see Table 3 in the Appendix.

For illustration purposes only (because many intermediate values are not given, the given values are rounded off and linear interpolation is crude!), Table 5.8 gives selected values of the working probits.

Remark 2: Fortunately, there are computer programs which will do all these steps for you! Indeed, at the Institute of Computer Science, University of Guelph, there are on file two such programs: *PROBITANAL*, an *APL* program by Prof. Jack Douglas, and the BMD03S program, a FORTRAN program out of U.C.L.A. (An investigation of these programs can be found in Malcolm, Pursey and Hubert (1976).) See the subsections 5.10.4 and 5.10.5, where sample outputs and a short discussion of these two programs are given.

Table 5.8

y, THE WORKING PROBITS

Y(Expected Probits)

p (%)	2.0	2.5	3.0	3.5	4.0	4.5	5.0	5.5	6.0	6.5	7.0	7.5
0	1.70	2.15	2.58	2.98	3.34	3.62	3.75	3.54	2.52	—		
5	—	5.00	3.51	3.37	3.55	3.77	3.87	3.68	2.73	—		
10	—	7.85	4.43	3.76	3.76	3.91	4.00	3.82	2.94	0.07		
15		—	5.36	4.14	3.96	4.05	4.12	3.96	3.14	0.45		
20		—	6.28	4.53	4.17	4.19	4.25	4.10	3.35	0.84		
25			7.21	4.91	4.38	4.33	4.37	4.25	3.56	1.23		
30			8.14	5.30	4.58	4.48	4.50	4.39	3.76	1.61		
35			9.06	5.69	4.79	4.62	4.62	4.53	3.97	2.00		
40			9.99	6.07	5.00	4.76	4.75	4.67	4.18	2.38		
45			—	6.46	5.20	4.90	4.88	4.81	4.38	2.77		
50			—	6.85	5.41	5.04	5.00	4.96	4.59	3.16	—	
55				7.23	5.62	5.19	5.13	5.10	4.80	3.54	—	
60				7.62	5.82	5.33	5.25	5.24	5.00	3.93	0.01	
65				8.00	6.03	5.47	5.38	5.38	5.21	4.31	0.94	
70				8.39	6.24	5.61	5.50	5.52	5.42	4.70	1.86	
75				8.78	6.44	5.75	5.63	5.67	5.62	5.09	2.79	
80				9.16	6.65	5.90	5.75	5.81	5.83	5.47	3.72	—
85				9.55	6.86	6.04	5.88	5.95	6.04	5.86	4.64	—
90				9.93	7.06	6.18	6.00	6.09	6.24	6.24	5.57	0.58
95				—	7.27	6.32	6.13	6.23	6.45	6.63	6.50	4.26
100				—	7.48	6.46	6.25	6.38	6.66	7.02	7.42	7.85

5.10.3 *Example*

Here is an example of this method. The dose (in milligrams of dry root per litre of spray fluid) of the insecticide rotenone was applied under standardized conditions to samples (roughly of size 50) of the insect species *Macrosiphoniella sanborni*. (This experiment is reported in Martin (1942) and is analyzed in great detail in Finney (1971).) Here, n = number of insects tested (at each X value), r = number of insects responding (dead or "badly affected" out of n).

Table 5.9

TOXICITY OF ROTENONE

dose	n	r	X	p(%)	y	Y'	y'	w	nw
0	49	0	$-\infty$	0.0	$-\infty$	—	—	—	—
2.6	50	6	0.415	12.0	3.825	3.9	3.828	0.3800	19.05
3.8	48	16	0.580	33.3	4.570	4.6	4.560	0.5932	28.47
5.1	46	24	0.708	52.2	5.055	5.1	5.050	0.6360	29.26
7.7	49	42	0.886	85.7	6.067	5.8	6.048	0.4227	20.71
10.2	50	44	1.009	88.0	6.175	6.3	6.165	0.3823	19.12

The variable $X = \log_{10}$ (dose) to three decimal places, and $p = 100 \cdot r/n$, to one decimal place. We let the probit of p be y, and probit(0) $= -\infty$ and probit(100) $= +\infty$.

We are now ready to proceed through the steps of the maximum likelihood procedure to estimate LC50. Under the normal assumptions we first find the plot of y versus X (use semi-log paper and ignore any $-\infty$ and $+\infty$ points) and by eye only draw in a provisional straight line. Suppose this was done and suppose further that this line generated the set of expected probits Y_i'. (Notice that we ignore the $-\infty$ value because $X = 0$; if $X \neq 0$ and p = 0% or 100%, we can get Y' values and they should be used in subsequent steps!) Then find the working probits y' and their associated weighting coefficients w; then compute the totals Σnw, ΣnwX, $\Sigma nwy'$, ΣnwX^2, $\Sigma nwy'^2$, $\Sigma nwXy'$, so that the first regression line can be found.

For this example:

$\Sigma nw = 116.61$, $\Sigma nwX = 82.7963$, $\Sigma nwX^2 = 63.2849$, $\overline{X} = 0.7100$,

$\Sigma nwy' = 593.6385$, $\overline{y}' = 5.0908$, $\Sigma nwy'^2 = 3101.5826$, $\Sigma nwXy' = 440.2130$,

$S_{XX} = \Sigma nwX^2 - ((\Sigma nwX)^2/\Sigma nw) = 4.49725$

$S_{y'y'} = \Sigma nwy'^2 - ((\Sigma nwy')^2/\Sigma nw) = 79.48614$

$S_{Xy'} = \Sigma nwXy' - ((\Sigma nwX)(\Sigma nwy')/\Sigma nw) = 18.71337.$

Thus the first iteration yields the estimates

$\quad b = S_{Xy'}/S_{XX} = 4.1611,$ $\qquad a = \overline{y}' - b\overline{X} = 2.1363$

and so we have $\qquad y' = 2.1363 + 4.1611X.$

The adequacy of this line can be tested by using the method described in section 4.7. By the way, there is an easier formula than (4.13) in the ML method. Since

$$\chi^2 = \sum \frac{(r - nP)^2}{nP(1 - P)}$$

$$= \sum \frac{n(p - P)^2}{P(1 - P)} \text{ , since } r = np$$

$$= \sum nw \left(\frac{p - P}{f} \right)^2 \text{ , by (5.29)}$$

$$= \sum nw (y - Y)^2, \text{ by (5.33),}$$

and since we can associate this y with y' and this Y with a + bX, then it follows that

$$\chi^2 = S_{y'y'} - \frac{(S_{Xy'})^2}{S_{XX}} , \tag{5.34}$$

which possesses (asymptotically) an approximate χ^2 distribution with k $-$ 2 degrees of freedom where k is the number of doses compared. If the fit is satisfactory then complete the analysis by step (7). If the fit is poor use the iterative argument suggested in steps (3) to (6).

For our example

$$\chi^2 = 79.486 - \frac{(18.713)^2}{4.497} = 1.62$$

which is less than $\chi^2_{.95}(3) = 7.8$, so that after the first iteration we have found the probit line. Therefore the ML estimate of log LC50 is

$$m = \frac{5 - 2.1363}{4.1611} = 0.6882$$

with

$$SE(m) = \frac{1}{b} \sqrt{\frac{1}{\Sigma nw} + \frac{(m - \overline{X})^2}{S_{XX}}}$$

$$= \frac{1}{4.1611} \sqrt{\frac{1}{116.61} + \frac{(0.6882 - 0.71103)^2}{4.49725}}$$

$$= 0.02239.$$

Thus LC50 $= 10^{0.6882} = 4.878$ mg/ℓ. The approximate 95% confidence interval is $(10^{m - 1.96SE(m)}, 10^{m + 1.96SE(m)}) = (4.41, 5.40)$ mg/ℓ. (Because the above calculations were done by hand calculator, they differ slightly from Finney's (1971) results: y' $= 2.13 + 4.18X$, $\chi^2 = 1.67$, m $= 0.686$, SE(m) $= 0.022$, LC50 $= 4.85 \pm 0.25$ mg/ℓ.

5.10.4   *An APL Program*

At the University of Guelph there are two computer programs which are based on the maximum likelihood procedure: an *APL* program and a FOR-TRAN program.

An *APL* program, *PROBITANAL,* was conceived by Jack Douglas and is stored in the *APL* public libraries. (The last revision was in January 1976.) It can be retrieved by typing)*LOAD 28 PROBIT* and then simply *PROBITANAL.* It utilizes a very simple conversational type input. To understand this program it suffices to study the following pages where a typical output is given for the example in Table 4.1.

PROBITANAL

IF YOU WISH INFORMATION CONCERNING CONSTRAINTS ON DATA FORM TYPE 1--
IF YOU ARE FAMILIAR WITH REQUIRED DATA FORM BEGIN BY TYPING 2
□:

1

NOTE WITH REGARD TO FORM OF DATA PRESENTED
1) ZERO DOSE SHOULD BE SPECIFIED ONLY WHEN RESPONSES ARE OBSERVED IN THE UNTREATED GROUP--THE
   ACTUAL RESPONSES WILL THEN BE ADJUSTED ON THE BASIS OF ABBOTT'S FORMULA
2) THE DOSE METAMETER USED IN THIS ANALYSIS IS LOG DOSE--THE USER MUST ENSURE THAT, WHERE NECESSARY,
   A TRANSFORMATION HAS BEEN APPLIED SO THAT FOR TREATED GROUPS THE ACTUAL DOSE VALUES SUBMITTED
   ARE GREATER THAN 1.0
3) IF YOU HAVE A SERIES OF SUCCESSIVE 0% EFFECTS AT THE LOW DOSAGES, OR A SERIES OF SUCCESSIVE
   100% EFFECTS AT THE HIGH DOSAGES, USE ONLY ONE OF SUCH EFFECTS, IN EITHER CASE THE ONE NEAREST
   THE MIDDLE OF THE RANGE OF DATA
4) IF % MORTALITY AT AN EXPERIMENTAL DOSAGE IS SMALLER THAN % MORTALITY IN THE CONTROL, ABBOTT'S
   FORMULA CANNOT ADEQUATELY ALLOW FOR THIS AND CALCULATIONS WILL BE BIASED
IF DATA SATISFY THE ABOVE CONSTRAINTS TYPE 1--OTHERWISE EXIT BY TYPING 2
□:

1

GIVE (IN ORDER OF INCREASING MAGNITUDE) DOSES FOR WHICH RESPONSE DATA WILL BE SUPPLIED
□:
    10 15 20 30 50 70 95
GIVE VECTOR OF THE NUMBER OF ANIMALS USED AT EACH DOSE LEVEL
□:
    47 53 55 52 46 54 52
GIVE VECTOR OF RESPONSES OBSERVED AT EACH DOSE LEVEL
□:
    8 14 24 32 38 50 50

85

DOSE	LOG DOSE	NUMBER HOSTS	RESPONSE PERCENT RESP	ADJ RESP	EMP PROBIT	EXPEC PROBIT	WT FACT	WRKG PROBIT	NWX	NWY
10.00	1.000	47	8.0 17.02	17.021	4.050	3.960	8.419	4.051	8.419	34.103
15.00	1.176	53	14.0 26.42	26.415	4.370	4.466	17.976	4.372	21.141	78.594
20.00	1.301	55	24.0 43.64	43.636	4.840	4.825	23.455	4.840	30.515	113.521
30.00	1.477	52	32.0 61.54	61.538	5.290	5.331	23.995	5.293	35.443	127.002
50.00	1.699	46	38.0 82.61	82.609	5.940	5.969	16.583	5.938	28.174	98.471
70.00	1.845	54	50.0 92.59	92.593	6.450	6.389	13.508	6.443	24.923	87.035
95.00	1.978	52	50.0 96.15	96.154	6.770	6.770	8.078	6.768	15.976	54.672

MEAN OF x= 1.469399179   MEAN OF Y= 5.297611297

SLOPE OF PROBIT LINE = 2.898834254
       VARIANCE 0.1119122786 STANDARD ERROR 0.3345329261

ED50 VALUE 23.26665041
       VARIANCE 0.001202277055 STANDARD ERROR 0.03468100424

LIMITS OF ED50 ARE 19.89569416 AND 27.20875267

***   ADJUSTED PROBIT VALUES   ***
   3.937   4.447   4.810   5.320   5.963   6.387   6.771

IF OUTPUT OF SUMS OF SQUARES IS DESIRED TYPE 1
----OTHERWISE IF NOT DESIRED TYPE 2
□:
     1

SNWX	SNWY	
164.59075	593.39731	
SNWXX	SNWYY	SNWXY
250.78518	3218.98000	897.84057
(SNWX$_x$SNWX)/SNW	(SNWX$_x$SNWY)/SNW	(SNWX$_x$SNWY)/SNW
241.84961	3143.58828	871.93784
S(XX)	S(YY)	S(XY)
8.93557	75.39172	25.90274

```
IF OUTPUT OF CROSS PRODUCTS IS DESIRED TYPE 1
------OTHERWISE IF NOT DESIRED TYPE 2
□:
     1
CALCULATED CHI SQUARE 0.303978934

     DEGREES OF FREEDOM 5
CHISQUARE BY METHOD OF GOLDSTEIN 0.08143776191

     DEGREES OF FREEDOM 5

IF PLOT OF OBSERVED (EMPIRICAL) AND ADJUSTED PROBITS VERSUS
LOG OF DOSE IS DESIRED TYPE 1--OTHERWISE TYPE 2
□:
     1
```

```
7.0 |
    |
    |                                                          *
    |
6.5 |                                                 □
    |                                                 *
    |
    |
    |
6.0 |                                        *
    |                                        □
    |
    |
5.5 |
    |
    |                            *
    |
5.0 |
    |
    |                *
    |
4.5 |
    |        *
    |
    |
    □
4.0 |
  * |
    |
    |
3.5 |
    1.0        1.2        1.4        1.6        1.8        2.0
```

IF ANOTHER CYCLE OF CALCULATIONS IS DESIRED USING THE WORKING PROBITS
DERIVED ABOVE TYPE 1----OTHERWISE EXIT BY TYPING 2
□: 1

DOSE	LOG DOSE	NUMBER HOSTS	RESPONSE	PERCENT RESP	ADJ RESP	EMP PROBIT	EXPEC PROBIT	WT FACT	WRKG PROBIT	NWX	NWY
10.00	1.000	47	8.0	17.12	17.120	4.050	3.961	8.396	4.055	8.396	34.045
15.00	1.176	53	14.0	26.50	26.500	4.370	4.466	17.924	4.375	21.081	78.412
20.00	1.301	55	24.0	43.64	43.637	4.840	4.825	23.396	4.840	30.439	113.238
30.00	1.477	52	32.0	61.53	61.530	5.290	5.330	23.956	5.293	35.386	126.794
50.00	1.699	46	38.0	82.60	82.597	5.940	5.966	16.586	5.938	28.179	98.483
70.00	1.845	54	50.0	92.56	92.561	6.440	6.386	13.533	6.441	24.970	87.169
95.00	1.978	52	50.0	96.15	96.154	6.770	6.766	8.109	6.768	16.038	54.885

MEAN OF X= 1.469951728 MEAN of Y= 5.299559241

SLOPE OF PROBIT LINE = 2.894159329
VARIANCE 0.1119031624 STANDARD ERROR 0.3345193004

ED50 VALUE 23.25128596

VARIANCE 0.001210019457 STANDARD ERROR 0.03478533393

LIMITS OF ED50 ARE 19.87319633 AND 27.20359071

*** ADJUSTED PROBIT VALUES ***
3.939   4.449   4.811   5.320   5.962   6.385   6.769

IF OUTPUT OF SUMS OF SQUARES IS DESIRED TYPE 1
----OTHERWISE IF NOT DESIRED TYPE 2
□: 2

IF OUTPUT OF CROSS PRODUCTS IS DESIRED TYPE 1
----OTHERWISE IF NOT DESIRED TYPE 2
□: 2

IF PLOT OF OBSERVED(EMPIRICAL) AND ADJUSTED PROBITS VERSUS
LOG OF DOSE IS DESIRED TYPE 1--OTHERWISE TYPE 2
□: 2

IF ANOTHER CYCLE OF CALCULATIONS IS DESIRED USING THE WORKING PROBITS
DERIVED ABOVE TYPE 1----OTHERWISE EXIT BY TYPING 2
□: 2

88

5.10.5   *A FORTRAN Program*

The second program is a Fortran program, commonly referred to as the BMD03S program, which is part of the package of programs called Biomedical Computer Programs (or simply, BMD), and was last revised in June 1972. A detailed description of the program, the theoretical basis of the program and the information necessary to invoke the program is available in the BMD manual.

In order to run this program the user must have a user number and project number and these numbers must be entered on the JOB card according to the usual job procedure.

The input consists of three items in this order:

(1) d_i, the ith dose level
(2) r_i, the number of subjects responding at the ith dose level, and
(3) n_i, the number of subjects tested at the ith dose.

The output may not at first be too clear: the estimate of LC50 is listed as MU-HAT. The output includes an optional listing of the input data, namely the untransformed dose levels, the responses, the sample sizes, as well as the corresponding probits and proportions of response. (The printout is horizontal and not the familiar vertical-type of table.) The printout also lists the coefficients of the fitted probit line, called *ALPHA* and *BETA,* and the chi-square value for testing the adequacy of the probit line. SIGMA-HAT is $\hat{\sigma} = 1/\hat{b}$. There is no interval estimate (e.g. 95% confidence interval); it can be obtained manually by the user from the COVARIANCE MATRIX and there is no plot of probit versus dose (for any iteration) and in fact there is no option to obtain it, as in PROBITANAL.

The BMD03S program utilizes Abbott's formula to adjust observed response rates with a control response rate C. In general this quantity C is a parameter and must be included in the likelihood function which is maximized. One feature of this program is that this is done. (The BMD reference manual refers to this as "optimizing" the value of C.) Since the iterative nature of the maximum likelihood method requires a starting value, say C_0, the choice of C_0 can be (a) inputted on a card (assuming prior knowledge), (b) set equal to an observed experimental value, or (c) defined as the minimum of all the observed response rates.

The following pages provide a copy of the 12 cards used to initiate the BMD03S program and a copy of a computer output. The above comments should clarify this output.

There are other excellent programs for probit analysis if these two programs are not available; for example see Russell and Robertson (1979) or Robertson et al (1981a), (1981b) and for the SAS system see Barr et al (1976); also see Feder and Sherrill (1980).

```
//
/*
FINISH
4700 5300 5500 5200 4600 5400 5200
800 1400 2400 3200 3800 5000 5000
10 15 20 30 50 70 95
(7F6.2)
SPECTG2 9100.00 3
PROBLEMTESF39      7
//SYSIN 00 *
// EXEC BMDLOAD,PROGRAM=BMD03S
//JOBNAME JOB (PROJ,USER), 'PROGRAMMER', NAME CLASS
```

A sample card deck for the BMD03S program.

BMD03S BIOLOGICAL ASSAY, PROBIT ANALYSIS-REVISED JUNE 2, 1972
HEALTH SCIENCES COMPUTING FACILITY, UCLA.

PROBLEM TESP39.
CONTROL GROUP DATA. SAMPLE SIZE N= 0., RESPONSE R= -0:. C HELD CONSTANT.
UNTRANSFORMED DOSE.

	0.10	0.15	0.20	0.30	0.50	0.70	0.95
RESPONSE.	8.00	14.00	24.00	32.00	38.00	50.00	50.00
SAMPLE SIZE.	47.00	53.00	55.00	52.00	46.00	54.00	52.00
SAMPLE PROBIT	4.05	4.37	4.84	5.29	5.94	6.45	6.77
SAMPLE PROBABILITY	0.17	0.26	0.44	0.62	0.83	0.93	0.96

TRANSFORMED DOSE FOR TRANSFORMATION CARD NO 1 OF 1
1.00   1.18   1.30   1.48   1.70   1.85   1.98

THE INITIAL VALUES AT START OF ITERATION ARE, ALPHA=
1.087 BETA= 2.873 C= -0.0

THE FINAL VALUES OF THE PARAMETERS AND THE PARTIAL DERIVATIVES ARE

FOR PROBIT LINE (ALPHA1+BETA*(X-XBAR))

		* FOR PROBIT LINE (ALPHA+ BETA*(X))
ALPHA1 = 5.378	DL/DA = 0.3397E-05	* ALPHA = 1.076
BETA = 2.875	DL/DB = 0.4388E-05	* BETA = 2.875
C = -0.0	DL/DC = 0.27330E 01	* C = -0.0

THE ESTIMATES MU-HAT AND SIGMA-HAT OF THE MEAN AND STANDARD DEVIATION OF THE TRANSFORMED DOSE
TOLERANCE ARE
    MU-HAT = 1.3650
    SIGMA-HAT = 0.3478

THE COVARIANCE MATRICES ARE AS FOLLOWS   *   FOR ALPHA, BETA.
    FOR ALPHA1, BETA.   *   0.1544 −0.1044
        0.0065   0.0056   *   −0.1044   0.0735
        0.0056   0.0734
    FOR MU, SIGMA.
        0.0011 −0.0006
       −0.0006   0.0011

CHI-SQUARE= 0.5451, WITH 5 DEGREES OF FREEDOM.

END FOR OUTPUT USING TRANSFORMATION CARD NO 1 OF 1

    END OF OUTPUT FOR PROBLEM TESP39

NO NEW PROBLEMS, PROGRAM TERMINATED BY FINISH CARD.

5.11   *Serial Dilution Assays*

To estimate the density of a specific organism in a suspension, a common method is to form a series of dilutions of the original suspension. Then from each dilution a specified volume, a dose, say, is placed in each of several tubes. Later the tubes are examined for evidence of growth of the organism.

Specifically, suppose θ represents the density of an organism, and in fact suppose it is known that θ is somewhere between 10^2 and 10^5 organisms per unit volume. Of course a serial dilution assay permits one to estimate θ. Suppose a ten-fold dilution factor is used. Then to assure doses span the range .01 to 5 organisms per dose, it is necessary to use concentrations ranging from 10^{-1} to 10^{-7} dilutions of the original suspension or 7 dilutions altogether.

As an example, consider the following data, where n = 3 tubes were used at each dilution and r represents the number of tubes showing growth. Let p = r/3 be the proportion of tubes showing growth.

Dilution	r
10^{-7}	0
10^{-6}	0
10^{-5}	0
10^{-4}	0
10^{-3}	2
10^{-2}	3
10^{-1}	3

The relevant estimators for θ are given by the following result:

Theorem

(a) A point estimator of θ (often referred to as the *most probable number*) is

$$\hat{\theta} = \exp\left(-\gamma - x_0 - \frac{d}{2} + d \sum_{i=1}^{k} p_i \right) \tag{5.35}$$

where $x_0 = \ln(z_0)$, z_0 is defined by $z_i = a^{-i} z_0$ (i = 0, 1, 2, . . ., k), the concentrations in the k + 1 dilutions, a > 1 is the dilution factor (i.e., z_0 is the highest concentration of the original suspension), the dose levels are equally-spaced d = ln (a) units apart, $\gamma = 0.57722$ is Euler's constant, and p_i are the observed proportions of tubes showing growth.

(b) A point estimator of the number of organisms per z_0 volume is

$$z_0 \hat{\theta} = \exp\left(-\gamma - \frac{d}{2} + d \Sigma p_i \right).$$

(c) The 95% confidence interval estimator is

$$\hat{\theta} \ e^{-1.96(d\ln2/n)^{1/2}} < \theta < \hat{\theta} \ e^{1.96(d\ln2/n)^{1/2}}.$$

Proof: We prove (a) only, because (b) follows directly from (a) and the proof of (c) can be found in Johnson and Brown (1961). Also see Robson et al (1961).

Proof of (a): Let r_i be the number of tubes showing growth out of n_i tubes at dilution z_i and θ be the density of organisms per unit volume in the original suspension. It is known from empirical evidence that the distribution of organisms in the individual doses is Poisson. Then the probability of growth (i.e., the probability of one or more organisms) in a tube receiving a dose of concentration z_i is

$$P(z_i) = 1 - P(\text{no growth}) = 1 - e^{-\theta z_i}.$$

If the experiment is constructed so that the dose levels are equally spaced on the log scale, let $x = \ln z$, $d = \ln a$ where a is the dilution factor. Then we write the probability of a growth response at dose level x as

$$F(x) = 1 - e^{-\theta e^x}$$

where F is called the *dose-response function* and has all the characteristics of a distribution function. The probability function is

$$f(x) = \theta \ e^{x - \theta e^x} \ ;$$

and so

$$\mu = \int_{-\infty}^{\infty} x \cdot f(x) \ dx$$

$$= \int_{-\infty}^{\infty} x \ \theta \ e^x \ e^{-\theta e^x} \ dx$$

$$= \int_{0}^{\infty} (\ln y - \ln \theta) \ e^{-y} \ dy$$

$$= - (\gamma + \ln \theta),$$

since the Euler constant $\gamma = \int_{0}^{\infty} e^{-y} \ln y \ dy = 0.57722$. Therefore

$$\theta = e^{-\gamma - \mu}.$$

Since μ is the mean of $F(x)$ and since this is exactly the same situation as a toxicity experiment of this chapter, then μ is estimated by the median of F or the LC50. If we take the Spearman-Kärber estimator we have

$$\hat{\mu} = x_0 + \frac{d}{2} - d \sum_{i=0}^{k} p_i$$

where $x_0 = \ln(z_0)$ and p_i is the observed proportion of tubes showing growth. Thus the estimator for θ is

$$\hat{\theta} = \exp\left(-\gamma - x_0 - \frac{d}{2} + d\,\Sigma p_i \right).$$

By the way, it can be shown that the distribution of $\hat{\mu}$, given x_0, is asymptotically normal (since it is a linear function of the sum of binomially distributed random variables) with $E[\hat{\mu}] \simeq \mu$ and $V[\hat{\mu}] \simeq (d \ln 2)/n$.

Example: For the data above, we have n = 3, $\Sigma p_i = 8/3$, $z_0 = 10^{-1}$, $x_0 = \ln 10^{-1}$, d = ln 10. Then

$$\hat{\theta} = \exp\left(-0.57722 - \ln 10^{-1} - \frac{\ln 10}{2} + (\ln 10)(8/3) \right)$$

$$= \exp(-0.57722 + 7.2915)$$

$$= \exp(6.71428)$$

$$\simeq 824 \text{ organisms per unit volume of the original suspension.}$$

The estimate of the number of organizms per 10^{-1} volume is $z_0\,\hat{\theta} \simeq 82$. And the 95% confidence interval for θ is

$$(824.1\ e^{-1.43},\ 824.1\ e^{+1.43}) = (197,\ 3442),$$

since $1.96((\ln 10 \cdot \ln 2)/3)^{1/2} = 1.43$.

It is important to remark here that $\hat{\theta}$ is a biased estimator of θ. In fact, it can be shown that

$$E[\hat{\theta}] \simeq \theta\left(1 + \frac{d \ln 2}{n} \right).$$

Also the standard error of $\hat{\theta}$ is

$$SE(\hat{\theta}) \simeq \left(\frac{\hat{\theta} \cdot d \cdot \ln 2}{n} \right)^{1/2}.$$

The coefficient of variation of $\hat{\theta}$, $CV(\hat{\theta})$ is given by

$$CV(\hat{\theta}) \simeq \left(\frac{d \ln 2}{n} \right)^{1/2}.$$

5.12   *Exercises*

5.1   Construct a scatter diagram of Y versus X for the data of Table 4.4 and then superimpose the estimated probit line.

5.2   As an exercise in technique find the point *and* interval estimate (95% confidence limits) for LC50 by probit analysis and by the Reed-Muench method for the following data. (Hint: use the dose metameter $X = \log_{10} d$.) (Data slightly changed from Martin (1942).)

dose (mg/ℓ)	number of insects (n)	number affected (r)	percent killed (p)
2.6	100	12	12.0
3.8	96	32	33.3
5.1	92	48	52.2
7.7	98	84	85.7
10.2	100	88	88.0

5.3   This strange data resulted from a study of the toxicity of ethylene oxide to the grain beetle, *Calandra granaria*. (Only the records of insects examined one hour after exposure to the poison were used.) Here X is $\log_{10}$ of the concentration of ethylene oxide in mg./100 ml. Estimate log LC50 by probit analysis, if possible! (Data slightly altered from Busvine (1938); also see Bliss (1940) and Finney (1971).)

X	n	r
.167	62	20
.199	60	34
.225	62	24
.260	54	14
.314	52	46
.322	60	44
.362	62	58
.391	60	46
.394	60	59

5.4 The following data concern a copper toxicity experiment to rainbow trout (courtesy of J. Sprague, Department of Zoology, University of Guelph). Different concentrations ($\mu g/\ell$ of CU^{++}) of copper sulphate were maintained in 5 different tanks, each containing n = 20 fish. The water hardness and the temperature was the same for each tank. The data below give the mortality in each tank after 24 hours. For those interested in the experimental design, see Sprague (1969), (1970), (1971), (1973), Hodson and Sprague (1975) and Eckart and Mahoney (1971). For a multivariate version of such an experiment see Chapter 7.

Concentration	270	410	610	940	1450
Mortality	5	6	15	19	20

(a) Find the probit line and test its adequacy.
(b) Find the point estimate of LC50.
(c) Find the weighted least squares estimate of the slope and its associated standard error.
(d) Find an approximate 95% confidence interval estimate of LC50.

5.5   The purpose of a recent study (see Daum and Givens (1963)) was to determine which of many treatments (all recommended for Boll weevil control) was the most toxic and which lost its effectiveness most rapidly. In this time-mortality experiment the insects were exposed for 4, 8, 24 and 48 hours to treated foliage. Below are the results for two such treatments, labelled here as A and B. Set 1 was from immediately after application; sets 2 and 3 were from 24 and 48 hours "weathering under artificial conditions". They concluded that treatment A lost $1 - .237 = 76\%$ of its effectiveness in 24 hours and 77% in 48 hours. ("Loss in toxicity or loss in effectiveness is $1 -$ potency".) Treatment B lost $1 - .513 = 49\%$ of its effectiveness in 24 hours and 75% in 48 hours. Also treatment A was the least toxic (LT50 = 13.8 hours) and treatment B was the most toxic (LT50 = 5.4 hours). Verify these conclusions.

Dose		Set 1			Set 2			Set 3	
		r			r			r	
(hours)	n	A	B	n	A	B	n	A	B
4	150	25	52	75	2	13			
8	200	80	143	175	10	71			
24	260	169	224	225	64	174	100	30	60
48	200	157	193	225	103	201	100	39	69

5.6   The results of ammonia as a fumigant using test insects at each of five concentrations are given below. Find the 95% confidence limits for LC50 using probit analysis.

log dose (μg/ℓ.) X	no. of subjects n	number responding r	percent responding p
.72	58	4	7
.80	60	14	23
.87	62	24	39
.93	56	38	68
.98	52	48	92

5.7   Discuss the toxicity of pyrethrum to *Tribolium castaneum* from the following data. (Slightly abridged from Tattersfield and Potter (1943), also see Finney (1971), p. 162.)

dose (μg/ℓ.)	EXPERIMENT 1 n	1 r	2 n	2 r	3 n	3 r	4 n	4 r
5	27	1	29	4	30	6	29	3
10	29	15	29	19	24	15	30	10
20	30	26	27	26	31	30	29	24
40	28	27	30	29	29	28	29	28

5.8   Discuss the following results of Atkins and Anderson (1954) on the toxicity of various pesticide dusts on honeybees, *Apis mellifera* L. These dusts were prepared with pyrophyllite as the diluent. Each material was evaluated by dusting at least 9 replicates of 20 bees each at each dosage level. The 5% DDT dust was included in each series of tests as the standard treatment, in addition to untreated checks.

Percent mortality at fixed hours after treatment for 3 levels of dose

Pesticide	%	400 mg. dosage							200 mg. dosage							100 mg. dosage						
		2	4	6	12	24	48	72	2	4	6	12	24	48	72	2	4	6	12	24	48	72
Aldrin	2	2	18	35	99	100			1	15	32	81	100			0	12	24	59	99	99	100
Parathion	2	21	39	66	72	100			26	50	66	73	98	100		22	44	57	73	100		
BHC	2	1	30	50	95	99	100		0	0	52	93	100			0	0	44	98	99	100	
Dieldrin	2	1	19	38	90	99	100		0	0	26	81	100			0	0	15	67	100		
Diazionon	5	0	24	47	71	99	100		19	31	54	74	100			0	6	36	62	98	100	
TEPP	1	36	36	38	40	98	100		37	37	37	45	96	100		28	28	29	47	95	100	
DDT	5	1	31	46	73	91	91	91	1	11	25	44	61	65	65	0	1	9	21	44	46	47
NPD	4	1	24	59	69	71	71	72	0	2	14	25	27	28	30	0	1	6	8	9	9	13
Sulfur	98	0	0	1	2	8	19	24	0	0	1	1	1	10	17	0	0	1	2	5	12	13
Demeton	1	0	0	0	2	5	10	13	0	1	2	4	5	8	9	0	3	3	4	4	9	11
Check	—	0	0	1	2	4	7	9	0	0	1	2	4	7	9	0	0	1	2	4	7	9

5.9   The following data is taken from a paper by Litchfield and Wilcoxon (1949) to illustrate a simple graphical procedure to obtain the relative potency of two drugs. The first set of values resulted from an experiment in which the response (actually the "antihistamine activity") of certain animals to Tagathen (Chlorothen citrate) was studied. They found ED50 = 0.18 mg/kg. The second set of values resulted from a similar experiment with tripelennamine (Pyribenzamine) as the agent. For this drug they found ED50 = 0.60 mg/kg. They then define the potency ratio as .60/.18 = 3.3. (Moreover, they suggest the 99% confidence limits are 1.6 to 6.8.) They conclude that "Tagathen was significantly more active than Pyribenzamine and its relative activity lies between 1.6 and 6.8 times that of Pyribenzamine".

     Using *any* method verify these conclusions. (The dose units are mg/kg and the number of animals tested at each dose was 8.)

Tagathen		Pyribenzamine	
Dose	No. Alive	Dose	No. Alive
0.025	1	0.175	1
0.125	4	0.35	3
0.25	4	0.7	5
0.5	7	1.4	5
1.0	8	2.8	8

5.10   Estimate the LC50 for the following data on the toxicity of ammonia to *Tribolium confusum*. Here $X = \log$ concentration in mg/l. (The data is slightly altered from Strand (1930); original data also analyzed in Finney (1971), page 176.)

X	n	r
.72	29	2
.80	30	7
.87	31	12
.93	31	18
.98	31	25
1.02	28	27
1.07	31	29
1.10	31	30

5.11 Data and pertinent calculations from a study of house fly mortality from pyrethrum spray are given below:

d	X	n	r	p	y	Y′	w	nw	y′
250	2.4	140	11	.079	3.59	3.6	.30	42.00	3.59
500	2.7	142	40	.282	4.42	4.7	.62	88.04	4.44
1000	3.0	134	110	.821	5.92	5.8	.50	67.00	5.91
2000	3.3	136	132	.971	6.90	6.9	.15	20.40	6.90

Source: Wadley (1967)

$\Sigma nw = 217.44$, $\Sigma nwX = 606.828$, $\Sigma nwy = 1078.408$

$\overline{X} = 2.7908$, $\overline{y} = 4.9596$, $\Sigma nwX = 3069.7146$

$\Sigma nwX^2 = 1708.888$, $\Sigma nwy^2 = 5588.3140$

$S_{XX} = \Sigma nw(X - \overline{X})^2 = \Sigma nwX^2 - (\overline{X})^2 \Sigma nw = 15.3642$

$S_{yy} = \Sigma nw(y - \overline{y})^2 = \Sigma nwy^2 - (\overline{y})^2 \Sigma nw = 239.8740$

$S_{Xy} = \Sigma nw(X - \overline{X})(y - \overline{y}) = \Sigma nwXy - \overline{X}\,\overline{y}\,\Sigma nw = 60.1151$

Notation: d = concentration in mg/ℓ , X = $\log_{10} d$, n = number of flies, r = number killed, p = mortality rate, y = probit of p, Y′ = expected probit from graph, w = weighting coefficient, y′ = working probit.

(1) Verify the values of p, y, Y′ w, nw, and y′ in the table.
(2) Construct a diagram of y versus X.
(3) By eye only, fit a line and read off the LC50 (the "quick method").
(4) Use probit paper to estimate LC50.
(5) Use an unweighted regression analysis to get a probit line. Test the adequacy of the fit and estimate LC50. (This requires some calculations.)
(6) Use a weighted linear regression to get a probit curve. (Here you can use the calculations provided.) Test the adequacy of the fit and estimate LC50.
(7) Perform a full scale maximum likelihood procedure to estimate LC50. (This step will require the working probits and lots of calculations; it is permitted to use the APL program *PROBITANAL* or the Fortran program BMD03S.)

5.12 A toxicity experiment on an insecticide using standard test insects in a standard manner was performed in a laboratory. On the first day, three concentrations were used on an adequate sample of insects with the results:

concentration units: 1   2   4
percent mortality: 25   42   73

There is special plotting paper available (with logarithmic intervals on the horizontal axis and probability intervals on the vertical axis) on which these readings can be plotted. The probit-log transformation relation should appear. A straight line can be fitted by eye, and the concentration for 50% mortality read off. (This is a version of the Litchfield-Wilcoxon method using "probit paper".) Estimate m = log LC50 (and LC50) for this data using this crude graphical procedure. Compare your answer with some other method of analysis.

5.13 (Continuation of 5.12) Suppose that instead of just one insecticide, three insecticides were analyzed in the laboratory and the object of the experiment was to compare their toxicities. The biologist decides that the best comparison of insecticides is that of strengths required for a given effect, i.e. LC50's. Suppose that the insecticides were each tested at three concentrations on six different days. Suppose further that for each day and for each insecticide an LC50 was obtained (as in exercise 5.12) and the 18 resulting LC50 values obtained were as follows:

Insecticide	Day					
	1	2	3	4	5	6
A	3.6	2.5	1.3	2.2	2.2	1.6
B	3.2	1.8	1.3	2.2	2.1	1.3
C	2.2	1.6	1.5	1.3	1.8	1.2

Consider the days as blocks in a randomized complete block design. (A fair assumption since days involved different groups of insects differing in susceptibility, and other uncontrolled variation.) Perform an ANOVA and verify the conclusions that ". . . this shows that the insecticides really differed" and a computation of the least significant difference indicates ". . . that A and B are not significant in their difference, but C is significantly lower than either". Why does it follow that C is "the most potent"?

5.14 The results of ammonia as a fumigant using 100 insects at each of five concentrations are given below. X = log dose ($\mu g/l$), p = percentage mortality rate, Y = probit of p.

X	p	Y
.72	14	3.92
.80	46	4.90
.87	44	4.85
.93	68	5.47
.98	92	6.41
4.30	—	25.55

Note:

$\Sigma X^2 = 3.74$, $\Sigma Y^2 = 133.9$

$\Sigma XY = 22.33$

Find the point estimate of the log median lethal concentration. (Note: you may assume the probit line adequately fits the data.)

5.15 Lukasewycz et al. (1975) studied the cellular immune response of C58/wm mice to syngeneic live malignant lymphoid cells.

Spleen Cells

| | Immune | | | Normal | |
X	n	r	X	n	r
3	36	1	4	18	0
4	36	7	5	34	1
5	44	33	6	34	6
6	26	25	7	46	33
7	25	25	7.7	10	8

Thymic Cells

	Immune			Normal	
3	10	0	5	10	0
4	10	0	6	26	2
5	36	11	7	15	12
6	36	23	8	16	16
7	26	25			

Bone Marrow Cells

	Immune			Normal	
3	10	0	5	18	0
4	20	1	7	18	0
5	36	5	7.5	6	5
6	36	29			
7	16	16			

For each of these six experiments find the log median survival "dose", where X is $\log_{10}$ and r is the number of survivors out of n mice. What are the implications of this data?

5.16 Plackett and Hewlett (1963) have studied the data of Turner (1955) for the toxicity to Milkweed bugs of two insecticides, here labelled for brevity by MOC and DFDT. The units for the dosage is "% concentration × 100", n is the number of bugs and r is the number of deaths. Find the 95% confidence limits for LC50 for each of the insecticides. Do the intervals overlap? (Hint: the authors suggest that slopes of the probit lines are 6.73 and 3.55 for MOC and DFDT respectively.)

MOC				*DFDT*		
d	n	r		d	n	r
5	48	9		20	48	5
7	48	22		28	29	7
10	48	42		40	48	18
14	48	47		56	48	28

5.17 White and Graca (1958) considered the toxicity of a cerium citrate solution to mice and obtained the following data:

d	50	100	150	200	250
n	30	30	30	30	30
r	1	5	20	19	25

The units of d were mg/kg and r is the number of deaths out of 30. Find the point and interval estimate for LC50.

5.18 The following data appears in Bartlett (1947). It is discussed in great detail in Bartlett (1937a) and (1937b). It represents the results of a fumigation experiment (24 hours exposure) on the bed bug. For the adults and the nymphs, sketch the plots of probit versus log dose on the same page, find the probit line, and the estimates of the log LC50; then find the point estimate of the relative potency. Bartlett (1947) claims the lines are: (adults) $Y' = -1.563 + 4.985X$; (nymphs) $Y' = -0.838 + 5.081X$. Comment on your analysis.

Dose	*Adults*		*Nymphs*	
(mg/ι)	Total	Dead	Total	Dead
7.83	20	2	10	0
11.76	25	3	9	5
17.20	24	6	5	3
19.00	23	6	11	10
20.90	24	19	7	6
23.20	22	8	10	5
24.60	13	8	18	17
28.00	9	9	27	24
29.80	28	21	4	3
32.00	25	19	3	3
36.90	15	14	17	17
40.20	17	17	15	15
44.90	8	8	20	20
Totals	253	140	156	128

5.19. In the following data, d = birth weight in ounces of infants, n = number of infants in the group, and r = number of "perinatal deaths" in the group. The problem was to determine the perinatal mortality-birth weight relationship for a population of human infants. The aim was to estimate the dose-response relationship for the 220 births weighing between 978 g (34.5 oz.) and 1772 g (62.5 oz.) which were born in the English counties of Devon and Somerset in 1965. (The data is from Fryer and Pethybridge (1975).) Here we treat perinatal death as quantal response and birth weight as stimulus.

d	n	r
34.5 − 38.5	15	8
38.5 − 42.5	24	18
42.5 − 46.5	21	13
46.5 − 50.5	32	19
50.5 − 54.5	39	23
54.5 − 58.5	43	15
58.5 − 62.5	46	12

Using probit analysis find the relationship by using d = midpoint of the interval. Then estimate the median birth weight. (It should be around 50 ounces.)

5.20 Irwin and Cheeseman (1939b) considered an experiment where the doses were in constant proportion and the number of subjects (n = 50) was the same for all doses of the toxin *Bacterium typhimurium*. Berkson (1953) has analyzed this data using "logits" and found LD50 = 0.227 mg. Use probits and estimate LD50.

Dose	X	r
0.0625	0	6
0.125	1	7
0.25	2	33
0.5	3	39
1.0	4	45
2.0	5	50

5.21 Graca et al. (1957) describe a comparative toxicity experiment of rare earth compounds using CFW albino mice weighing 20 ± 1.0 grams and raised and maintained in standard laboratory conditions. In the following data 5 dose levels in mg/kg were given to 30 animals each for each of four citrate complexes: C:cerium, N:neodymium, P:praseodymium, and L:lanthanum. The responses are numbers of mice dead after one week.

Dose	C	N	P	L
50	1	1	0	4
100	5	5	6	24
150	20	21	17	27
200	19	26	26	28
250	25	22	26	30

Find the point estimate of LD50 for each of the four rare earth compounds, using probit analysis. The published results are:

	LD50 ± SE	95% confidence limits
C	146.6 ± 9.7	128.6, 167.1
N	138.0 ± 16.5	94.4, 201.8
P	140.6 ± 7.6	126.2, 156.7
L	78.2 ± 9.9	70.0, 100.3

What are the implications of these results?

In this same paper they repeated the experiment using 9 guinea pigs per dose level:

Dose	C	P	N	L
25	0	0	1	0
50	6	5	8	4
100	7	9	8	9
150	8	8	7	7
LD50	55.7	53.0	40.5	60.7
SE	9.3	7.7	20.2	17.4
Lower limit	40.1	39.9	4.7	17.7
Upper limit	77.4	70.3	348.0	207.0

Compare these results with the corresponding data for mice.

5.22 Harris (1959) has discussed the results of the following toxicity experiment on the insecticide dieldrin in hard water (pH : 8.2, hardness: 400 ppm measured as calcium carbonate) to fathead minnows (*P. promelas*) exposed for 72 hours in groups of ten fish per dose.

Dose (ppm × 100)	log dose	% mortality
1.33	.125	0
1.78	.250	30
2.37	.375	50
3.16	.500	80
4.21	.625	100
5.62	.750	100

He found that the point estimate of LD50 is .023 ppm and the interval estimate (95%) is (.020, .028) ppm by using a "moving average-angle" method. Use some procedure discussed in this chapter (show all calculations) and verify these findings (use n = 100). Also use n = 10 and compare the interval estimates. (The Shuster-Dietrich method yields the answers: 0.024 and (0.0232, 0.0249) for the n = 100 case.)

5.23 Here is an interesting application of probit analysis, which was reported by Ashford and Smith (1965b). The data concerns the general problem of the prevalence of pneumoconiosis among coal miners. (Also see the paper by Ashford (1959).) A film on each of the 221 miners from a radiological survey was examined by two experienced radiologists, denoted here by A and B. The analysis assumes that the chance of developing pneumoconiosis is a function of the period spent mining. If "dose" is period of exposure (in years) and "response" is having pneumoconiosis, then we can cast this data into a quantal response type of assay.

d (years)	n	r A	r B
2.25	43	0	0
7.0	29	4	3
12.0	27	6	6
17.0	50	22	24
22.0	24	17	16
27.0	23	14	16
32.0	12	5	5
37.0	7	5	5
42.8	6	6	6

They claim that the point estimate of ED50 is 19 years for both A and B, and the 95% confidence limits are (16, 23) and (16, 22) for A and B respectively. Verify these results. (For A, the Shuster-Dietrich method yields the values 18.2 and (15.25, 21.71) for the point and interval estimates.)

5.24 In an assay of trypanosomes the experimental technique involved inspection under the microscope of a representative sample suspension. The number, r, of organisms observed to be dead out of n in the sample were recorded for each of the dose levels d (units omitted here). The data is slightly abridged from Ashford and Walker (1972).

d	n	r
4.7	55	0
4.8	49	8
4.9	62	18
5.0	55	18
5.1	53	22
5.2	53	37
5.3	51	47
5.4	50	50

Find the probit line and test for adequacy. If the fit is good, find LC50.

5.25 Plackett and Hewlett (1963) considered mixtures of drugs. Let A stand for "2.2% solution of 2 - n - Butyl - 4, 6 - Dinitrophenol", B for "2.2% solution of 2 - sec - Butyl - 4, 6 - Dinotrophenol" and C for A + B, with all solutions in Risella 17 oil. The toxicity to Lesser Mealworm beetles, *Alphitobius laevigatus,* of each drug applied singly and for the mixture is given below. Doses were varied by varying the volumes (d in μl) of the solutions. The control mortality was zero. Show that the sample probit lines are parallel. Also, sketch the lines.

For other references on the joint action of mixtures of drugs, see the fundamental papers by Ashford and the series of recent papers by Plackett and/or Hewlett. Also see Fowlkes (1979), Preston (1952), Sprague and Logan (1979) and Davis (1980).

A			B			C		
d	n	r	d	n	r	d	n	r
.0300	59	1	.0300	60	8	.0200	59	31
.0425	80	6	.0425	79	23	.0275	81	53
.0600	60	25	.0600	59	38	.0350	61	48

5.26 The following data is from Hewlett (1962) who studied the toxicity of DDT (1.4% DDT in Risella 17 oil) on two sites, mesoscutellum (M) and hind coxae (H), of the external surfaces of Lesser Mealworm beetles, *Alphitobius laevigatus* (F.). The volume of the solutions used was varied and the control response was zero.

Volume (μl)	M		H	
d	n	r	n	r
.025	60	4	60	1
.040	60	30	59	18
.065	59	51	60	45

For each of these sites find the probit line and then the estimate of LD50. Use the metameter $X = \log_{10} (100d)$. The authors suggest that DDT is estimated to be 1.198 times as potent when applied to the mesoscutellum of the beetle as when applied to its hind coxae. Suggest a biological reason for this.

5.27 Sibuya (1962) has analyzed the following data on a lethal poison, where X is the log dose and r is the number of animals dying out of n = 50.

X	−1.2	−0.8	−0.4	0	0.4	0.8	1.2	1.6
r	3	7	8	20	24	32	43	45

Estimate the log LC50 by some distribution-free method.

5.28 Female white rats (CFE albino strain) were used as subjects in a toxicity experiment on the insecticide Malathion. Probit analysis yielded the probit line $Y = 3 + 4X$, where $X = \log_{10}$ dose. Let LDp denote the level of Malathion which will result in a response of p% of the rats in the population. If LDp is estimated to be 4.121 mg/kg, then find the value of p.

5.29 The following data is given in Berkson (1960) and analyzed by Ashton (1972) using logit analysis which yields log ED50 = 1.99.

X($\log_{10}$ dose)	n	r	p
0	30	1	.0333
1	30	8	.2667
2	30	15	.5000
3	30	23	.7667

Find log ED50 using probit analysis.

5.30 The following data relates to a bioassay of a compound used to kill rats and mice. The substance was tested in various solvents such as water and glucose solution of varying strengths. The data below is only one set of such data for which the solvent was a 5% glucose solution. Unequal numbers of animals were tested at each of 5 dose levels, which were roughly in geometrical progression. The response was the number of dead animals.

The highest dose-level gave 100% response. (This data is analyzed by Ashton (1972) using logit analysis and yielded LC50 = 197.3. In order to apply this technique the p value for the highest dose level was replaced by $1 - 1/(2n) = 1 - (1/16) = .9375$.)

d (mg/kg)	X	n	r	p
98	1.9912	19	3	0.1579
137	2.1367	10	4	0.4000
192	2.2833	20	11	0.5500
268	2.4281	20	11	0.5500
375	2.5740	8	8	0.9375

Estimate LC50 by probit analysis.

5.31   In a larger serial dilution assay on infectious canine hepatitis, Robson et al (1961) conducted an experiment on 15 culture tubes per dilution level. After a fixed stage procedure which locates the approximate range which covers the median infection dose, ID50, a second procedure was designed with the results in the table below. The authors conclude that . . . "the range of dilutions over which infection changes from essentially 0 to 100% is approximately $10^{-3.1}$ to $10^{-3.7}$; that is, the change from neutralization to complete infection takes place over approximately a four-fold dilution range."

Using the Spearman-Karber estimator defined in equation (5.1), show that ID50 = $10^{-3.39}$.

Serum Dilution ($-\log_{10}$)	Number of Tubes Showing Lesions
2.6	0
2.7	1
2.9	1
3.0	0
3.1	0
3.2	9
3.3	0
3.4	6
3.6	12
3.7	15
3.8	15
3.9	15

5.32 The bibliography at the end of this text contains many theoretical and applied research papers. From the following list of recent papers (which are classified only here according to the type of subject used in the analysis), *or* from any other contemporary paper from your field of interest, choose a paper and write a two page essay on the techniques used and the conclusions reached. Fish: Lloyd (1960), Sprague (1964), Sprague and Ramsay (1965), Ballard (1969), Sanders (1970), Russo et al. (1974), McLeese (1974a), (1974b), Hodson and Sprague (1975), Wedemeyer and Nelson (1975), Davis and Hoos (1975), Sprague (1969), (1970), (1971), (1973), McCarty et al. (1978) and U.S. Vol. 3 EPA (1971) for recent references or Doudoroff and Katz (1953) for older references; for pre-1960 experiments in entomology, see Hoskins and Craig (1962) and a recent example is Vandenberg and Soper (1979); beetles: Bliss (1935a), Roussel et al. (1972), Vincent and Lindgren (1975); weevils: Brindley (1975), Thomas and Bradley (1975); chicks (or birds): Carpenter et al. (1963), Campbell (1966), Matterson et al. (1974); pheasants: Gill et al. (1970); rabbits: Young and Roman (1948); mice: Ozburn and Morrison (1964), Bazhanov (1971); poisons: Abbott (1925), Bliss (1939), Plackett (1952); protein: Hegsted (1968), Tomarelli (1962); menarche: Milicer and Szczotka (1966), Milicer (1968), David (1974), Jenicek (1974), Gallo (1977), Roberts et al (1977), Aranda-Ordaz (1981); drugs or medicine: the papers of Ashford et al., Hewlett et al. and Cranmer (1974); radioimmunoassays: Kemp (1959), Yalow and Berson (1968), Bliss (1970), Rodbard et al. (1970), (1971), (1974), Petrusz et al. (1971a), (1971b), Robyn et al. (1968), Healy (1972), Skelley, Brown and Besch (1973), Cook (1974), Finney et al. (1975), Gaines (1976), Tiede (1976), Tiede and Pagano (1977), Tsay et al (1978), Chapman (1979), Raab (1981). Other examples can be found in the U.S. Environmental Protection Agency data book, vol. 3, 1971.

6 Quantal Assays: Comparative Studies

6.1 *Example*

Suppose the following table represents the results of an experiment on mice. Here the experimenter recorded how many electric shocks could be applied to the tail of a mouse before the mouse squeaked; a specified dose of drug was then administered and a new shock-count was recorded. If the number of shocks increased by four or more, the mouse was considered as "responding". Large numbers (60 to 120) of (homogeneous) mice were tested at three dose levels for each of the drugs, morphine and amidone. The values of X represent the logarithm to the base 10 of the original data; n is the number of mice and r is the number responding. (This data is from Grewal (1952) and is analyzed in Finney (1971), p. 104.)

Table 6.1
A SIMPLE COMPARATIVE
QUANTAL ASSAY

Dose	Morphine		Amidone	
X	n	r	n	r
.18	103	19	60	14
.48	120	53	110	54
.78	123	83	100	81

The problem was to estimate the relative analgesic potency of amidone to morphine on mice. (Analgesia is the insensibility to pain without losing consciousness.)

6.2 *Analysis*

One obvious approach is to consider the two sets of data separately and use probit analysis. When we use the BMD03S program the provisional probit regression line for each drug is: $Y_1 = 3.74 + 2.23 X_1$ for morphine and $Y_2 = 3.73 + 2.71 X_2$ for amidone, where Y is the probit and X is the log dose. From these lines we can now estimate $m = \log_{10}$ ED50. In fact, if Y = 5 in each line we obtain $m_1 = 0.566$ for morphine and $m_2 = 0.470$ for amidone. (These answers slightly differ from Finney (1971) who used another calculation method and obtained 2.30 as the slope of both lines and $m_1 = .57$ and $m_2 = .49$ for morphine and amidone respectively.)

The next step is to realize that if two series of quantal response data yield parallel probit regression lines against the logarithm of the dose, then the difference between values of the dose which produce the same response rates is constant. (This is the key idea and one should sketch a diagram to illustrate this fact.) This implies a constant relative potency at all levels of response, where the relative potency of two agents is defined as the ratio of equally effective doses, so that log ρ is the difference. Thus the relative potency as a measure of the difference between the two agents can be found. This constant X-difference, which we shall denote as m, is equal to the difference of the log ED50's; for example, in the mice example $m = 0.566 - 0.470 = 0.096$. More formally, the amount by which a log dose in the first series exceeds an equally effective log dose in the second is called the *relative dose metameter* of the second series and is denoted by m, and due to parallelism of the regression lines, we must have

$$
\begin{aligned}
m = m_1 - m_2 &= (5 - \hat{a}_1)/\hat{b}_1 - (5 - \hat{a}_2)/\hat{b}_2 = (\hat{a}_2 - \hat{a}_1)/\hat{b} \\
&= ((\overline{Y}_2 - \hat{b}\,\overline{X}_2) - (\overline{Y}_1 - \hat{b}\,\overline{X}_1))/\hat{b} \\
&= (\overline{X}_1 - \overline{X}_2) - \frac{1}{\hat{b}}(\overline{Y}_1 - \overline{Y}_2)
\end{aligned} \tag{6.1}
$$

where the suffixes indicate the two stimuli and $\hat{b}$ is the estimate of the common slope of the probit regression lines. m is calculated from the data and is a statistic which estimates an unknown true value, namely, the log of the relative potency. If ρ is this true relative potency in the sense that a dose d of preparation 2 has the same average effect as a dose ρd of preparation 1, then the best estimate of ρ is r where

$$
r = 10^m. \tag{6.2}
$$

Much of the inferential problems parallel the discussions given in the section on the maximum likelihood approach. Chapter 6 of Finney (1971) should be consulted for more detail, examples, extensions and other procedures, including a very good illustration of a typical computer output of a probit analysis of 4 preparations!

6.3   *The Berkson Minimum Chi-Squared Method*

The chief competitor to the ML approach to estimate the log relative potency, which we have seen is given by

$$
m = \log \hat{\rho} = \frac{\hat{a}_t - \hat{a}_s}{\hat{b}},
$$

is due to Berkson (1949), (1955a). He called his approach the *minimum transform chi-squared method*. It is often simply referred to as the *minimum* χ^2 *method*. It is a non-iterative procedure with the estimates given by

$$\hat{a}_s = \overline{Y}_s - \hat{b}\ \overline{X}_s \tag{6.3}$$

$$\hat{a}_t = \overline{Y}_t - \hat{b}\ \overline{X}_t \tag{6.4}$$

$$\hat{b} = \frac{\displaystyle\sum_{i=1}^{k_s} n_{si}\ w_{si}\ (Y_{si} - \overline{Y}_s)(X_{si} - \overline{X}_s) + \sum_{j=1}^{k_t} n_{tj}\ w_{tj}\ (Y_{tj} - \overline{Y}_t)(X_{tj} - \overline{X}_t)}{\displaystyle\sum_{i=1}^{k_s} w_{si}\ n_{si}\ (X_{si} - \overline{X}_s)^2 + \sum_{j=1}^{k_t} w_{tj}\ n_{tj}\ (\overline{X}_{tj} - \overline{X}_t)^2} \tag{6.5}$$

where X_s and X_t are the log dose for the standard and test preparations, $Y_{ui} = \phi^{-1}(p_{ui})$ (u = s or t) where ϕ is the cumulative distribution function of the standard normal variable. The Y values are not probits, but are *normits*. Here $0 < p < 1$ and when p is zero or one, Y is undefined. For such extreme values Berkson suggests replacing the 0% rate by $1/(2n_{ui})$ and the 100% rate by $1 - 1/(2n_{ui})$. Brown (1964) has suggested a smaller transform: $1/(10n_{ui})$ and $1 - 1/(10n_{ui})$. The weights are defined by

$$w_{ui} = \frac{f_{ui}^2}{p_{ui}(1 - p_{ui})} \quad (u = s\ or\ t), \tag{6.6}$$

where f is the ordinate of the standard normal density function. Also

$$\overline{Y}_u = \frac{\displaystyle\sum_{i=1}^{k_u} w_{ui}\ n_{ui}\ Y_{ui}}{\displaystyle\sum_{i=1}^{k_u} w_{ui}\ n_{ui}}, \quad \overline{X}_u = \frac{\displaystyle\sum_{i=1}^{k_u} w_{ui}\ n_{ui}\ X_{ui}}{\displaystyle\sum_{i=1}^{k_u} w_{ui}\ n_{ui}}, \quad (u = s\ or\ t).$$

It is known that this minimum χ^2 method consistently gives smaller estimates than the maximum likelihood method (see Little (1968).) Other differences can be found in Cramner (1964), Gurland et al. (1960), Sowden (1972) and the excellent book by Hewlett and Plackett (1979); also see Vølund (1978), Berkson (1980) and Amemiya (1980).

6.4   *Quantit Analysis*

Recently Copenhaver and Mielke (1977a), (1977b) proposed a method which they call *quantit analysis* based on the omega function (see Prentice (1976)). It is a general procedure because under certain limiting conditions it includes methods based on the logistic distribution (logit analysis), the double exponential distribution, and the uniform distribution (linit analysis). It provides accurate estimates of extreme dosage levels (viz., LC5, LC99).

The *quantit* of p_i, $h_v(p_i)$, is defined by

$$h_v(p_i) = \int_{0.5}^{p_i} (1 - |2u - 1|^{v+1})^{-1} \, du \qquad (6.7)$$

where v is a shape parameter characterizing the distribution (e.g., v = 1 yields the logistic distribution). For further comments also see Tsutakawa (1980).

6.5 *Exercises*

6.1   Verify the calculations for the probit regression lines obtained in section 6.2 by using the method illustrated in chapter 4.

6.2   Verify that the estimate of log ED50 is 0.566 for morphine and 0.47 for amidone in the mice example by using the methods described in chapter 4.

6.3   The following results are from an assay of *Neoarsphenamine* by observing the proportion of negative blood smears in female rats injected with an agent.

	Dose (X)	No. of animals (n)	No. of negative smears (r)
Standard	12.00	8	0
	15.00	8	5
	18.75	7	6
Test	9.6	7	0
	12.0	8	6
	15.0	9	7

Estimate the relative potency. An interesting feature of this data is that doses are not in logarithmic increments and also the doses for standard and control are not the same. (This data is from Finney (1964), p. 461.)

6.4   The following data is taken from Salsburg (1971):

Organism	Dose	Exposed per Dose	Successes per Dose
A	1	62	55
A	2	38	38
A	3	27	26
B	1	25	7
B	2	5	5
B	3	42	39
C	1	62	0
C	2	38	0
C	3	49	1
Totals		348	

The experiment was as follows: an experimental drug was applied to a large group of patients (348) infected with one of 3 organisms (labelled here as A, B, C for simplicity). The study of protocols allowed the clinicians to choose any one of three different dose levels (labelled above as 1, 2, 3 for simplicity) for a given patient. Thus the sample sizes (see column 3) were random, although the doses chosen were not. Patients were recorded as cured (see column 4) if "all traces of the infection were eliminated". Otherwise the treatment was considered a failure. By the way, the drug had some history: it was known to be effective against two of the organisms (namely, A and B) and marginally effective against a third (labelled C above). This is an example of a quantal assay with the feature that the probabilities of response are near 0 or 1. (See Frawley (1974).) Analyze this data.

6.5   Gurland et al. (1960) have analyzed in great detail the following data which was taken from Pfaeffle (1958). The test preparation (T) is an extract prepared from alfalfa sprayed with a given amount of Guthion. The purpose of the bioassay is to estimate the amount of Guthion residue present in T. For comparison, the standard preparation (S) of Guthion is made to contain a given amount of control extract (10 ml in this case) in order that the masking effect due to plant lipids and other inactive substances present in T be the same for both preparations. Note that all doses of T contain the same total amount of plant extract (10 ml in this case), one part due to the test extract itself, and one part due to the control extract added. The 50 sub-

jects, houseflies, in every jar were observed at 17 hours of exposure and
the number dead (r) were counted. Estimate the log relative potency. (Answer: 1.4)

STANDARD PREPARATION
(c.e. = control extract)

Jar	Dose	r
1	20 μg + 10 ml c.e.	5
2	35 μg + 10 ml c.e.	21
3	45 μg + 10 m/ c.e.	35

TEST PREPARATION

Jar	Dose	r
1	1.0 ml + 9.0 ml c.e.	12
2	1.5 ml + 8.5 ml c.e.	28
3	2.0 m l + 8.0 ml c.e.	36

6.6   The data below are a result of a comparative assay of Rotenone relative to
Deguelin, using as a subject the chrysanthemum aphid, *Macrosiphoniella
sanborni*. The concentration, d, is in mg per litre. (The abridged data is
taken from Finney (1952, p. 69) but is originally due to Martin (1942); it is
also re-analyzed by Berkson (1953) using the minimum logit χ^2 method.)

	ROTENONE				DEGUELIN		
d	X	n	r	d	X	n	r
2.6	0.415	50	6	10.1	1.004	48	18
3.8	0.580	48	16	20.2	1.305	48	34
5.1	0.708	46	24	30.3	1.481	49	47
7.7	0.886	49	42	40.4	1.606	50	47
10.2	1.009	50	44	50.5	1.703	49	48

Show that the relative potency is 2.7 and interpret your answer.

6.7   This neurohormone bioassay was discussed by Prof. Gary Koch on April
13, 1977 at a seminar at McMaster University. Using the logit transform
and a weighted least squares procedure the estimate of ρ was found to be
0.2. Use probit analysis and estimate ρ. (The subjects were mice and r
represents the number of deaths out of n mice; the blanks in the data indicate that no subjects were used for that dose level for that drug.)

dose (mg)	.01	.03	.10	.30	1	3	10	30	100
n	30	30	10	10	10	10	10	10	10
drug A r	0	1	1	1	4	4	5	7	—
drug B r	—	—	—	0	0	1	4	5	8

6.8   The response proportions (p) were recorded for 9 different dosage (d) levels of a drug in a single agent quantal bioassay. Suppose X is the dose metameter log(d) to the base 10, and $Y = \text{probit}(p)$. Also a probit analysis yielded the mean of X to be 1/4, the mean of Y to be 4.5, and the slope of the probit line to be 14.52.

(a) Estimate LC50.

(b) Estimate LC60.

(c) Let LDα denote the level of the drug which will result in a response by α% of the experimental units in the population under study. If LDα is estimated to be 1.75, then find α.

7 A Multivariate Approach to Bioassays

In chapter 4 the method of probit analysis was used to analyze single agent quantal bioassays. The purpose of this chapter is to introduce another method and a new model that uses the correlations between observations over time to estimate LC50. This approach is based on multivariate growth curve theory and is outlined in section 7.1. A multivariate version of Fieller's theorem has been derived by Carter and Hubert (1984) and is used to construct simultaneous confidence intervals for the LC50 values at arbitrary time points in section 7.2. A procedure for testing the adequacy of the model is given in section 7.3. An example, courtesy of Professor John B. Sprague, Department of Zoology, University of Guelph, is used to illustrate this new methodology.

The basic reason for this approach is that when an experiment is monitored over a sequence of time points, the corresponding response variables become functions of concentration and time. Applying univariate probit analysis at each time point neglects the dependency in time and the possible interaction of time and concentration on the response. In fact, knowledge of these dependencies can be exploited to yield estimates which have statistical validity.

7.1 Notation and Data Structure

We need the following notation:

c_j = the j-th concentration level, with j = 1,2,...,d
d = the number of levels of concentration

$X_j = \log_{10}(c_j)$ is the j-th log concentration

T_i = i-th time point, with i = 1,2,...,t

n_{jk} = the number of sampling units at the j-th concentration and the k-th replication, with k = 1,2,...,r

z_{ijk} = the mortality count at the i-th time point for the j-th concentration and the k-th replication.

To understand this notation consider the data in Table 7.1 in which each of d = 7 different concentrations of a toxic copper substance was administered to tanks each containing 10 fish. The mortality counts for each tank were recorded after 48 hours (2 days) and again after 3 and 4 days; with these t = 3 time points the experiment had r = 2 replications.

Table 7.1
OBSERVED MORTALITY COUNTS

Block	Time (Days)	Concentration (μg/ℓ)						
		0.10	**0.20**	**0.30**	**0.50**	**1.00**	**2.00**	**2.50**
1	2	0	1	1	2	3	3	4
	3	1	1	2	3	5	6	7
	4	1	1	3	4	8	7	9
2	2	0	1	1	2	3	4	4
	3	1	1	2	3	5	6	7
	4	1	1	2	4	7	8	8

To illustrate one approach to the problem of estimating the LC50 for such an experiment, a variance-stabilizing transformation is first made on the mortality rates by letting

$$Y_{ijk} = \frac{1}{2}[\text{arcsine } (z_{ijk}/n_{jk} + 1)^{1/2} + \text{arcsine } (z_{ijk} + 1/n_{jk} + 1)^{1/2}].$$

$$(7.1)$$

Then the response variables Y_{ijk} are represented as a polynomial of order q-l in time and linear in log concentration; that is, we assume

$$Y_{ijk} = \sum_{m=1}^{q} (\beta_{m1} + \beta_{m2} X_j + \rho_{mk}) T_i^{-(m-1)} + \epsilon_{ijk} \qquad (7.2)$$

$i = 1,2,\ldots,t, j = 1,2,\ldots,d$ and $k = 1,2,\ldots,r$, with $\sum_{k=1}^{r} \rho_{ik} = 0$, for each i. (For our example $i = 1,2,3$, $j = 1,2,\ldots,7$ and $k = 1,2$.)

The error vector is denoted by

$$\grave{\epsilon}_{jk} = (\epsilon_{1jk},\ldots,\epsilon_{tjk})' , \qquad (7.3)$$

where the prime ' denotes the usual transpose of the vector. The $\grave{\epsilon}_{jk}$, j = 1,2,...,d, k = 1,2,...,r are assumed to be independently and normally distributed with mean zero and covariance matrix Σ.

This model can be cast into the form of the growth curve model. (See Potthoff and Roy (1964) for a discussion of this multivariate approach.) Using matrix notation we can write the model as

$$Y = B\beta A + \epsilon, \qquad (7.4)$$

where the $t \times dr$ response matrix (it is 3×14 in our example) is

$$
Y = \begin{bmatrix}
Y_{111} & \cdots & Y_{1d1} & & Y_{11r} & \cdots & Y_{1dr} \\
\cdot & \cdots & \cdot & & \cdot & \cdots & \cdot \\
\cdot & \cdots & \cdot & \cdots & \cdot & \cdots & \cdot \\
\cdot & \cdots & \cdot & & \cdot & \cdots & \cdot \\
Y_{t11} & \cdots & Y_{td1} & & Y_{t1r} & \cdots & Y_{tdr}
\end{bmatrix},
$$

the $t \times q$ design matrix for the time variable is

$$
B = \begin{bmatrix}
1 & T_1^{-1} & T_1^{-2} & \cdots & T_1^{-(q-1)} \\
1 & T_2^{-1} & T_2^{-2} & \cdots & T_2^{-(q-1)} \\
\cdot & \cdot & \cdot & \cdots & \cdot \\
1 & T_t^{-1} & T_t^{-2} & \cdots & T_t^{-(q-1)}
\end{bmatrix},
$$

the $q \times (r+1)$ parameter matrix is

$$
\beta = \begin{bmatrix}
\beta_{11} & \beta_{12} & \rho_{11} & \rho_{12} & \rho_{13} & \cdots & \rho_{1(r-1)} \\
\beta_{21} & \beta_{22} & \rho_{21} & \rho_{22} & \rho_{23} & \cdots & \rho_{2(r-1)} \\
\cdot & \cdot & \cdot & \cdot & \cdot & \cdots & \cdot \\
\beta_{q1} & \beta_{q2} & \rho_{q1} & \rho_{q2} & \rho_{q3} & \cdots & \rho_{q(r-1)}
\end{bmatrix},
$$

the $(r+1) \times dr$ design matrix for the concentration and block effects (it is 3×14 in our example) is given by

$$
A = \begin{bmatrix}
1 & \cdot\,\cdot\,1 & 1 & \cdot\,\cdot\,1 & & 1 & \cdot\,\cdot\,1 & & 1 & \cdot\,\cdot\,1 \\
X_{11} & \cdot\,.X_{d1} & X_{12} & \cdot\,.X_{d2} & & X_{1(r-1)} & \cdot\,.X_{d(r-1)} & & X_{1r} & \cdot\,.X_{dr} \\
1 & \cdot\,.1 & 0 & \cdot\,.0 & & 0 & \cdot\,.0 & & -1 & \cdot\,.-1 \\
0 & \cdot\,.0 & 1 & \cdot\,.1 & \cdots & 0 & \cdot\,.0 & & -1 & \cdot\,.-1 \\
\cdot & \cdots & \cdot & \cdots & & \cdot & \cdots & & \cdot & \cdots \\
0 & \cdot\,.0 & 0 & \cdot\,.0 & & 1 & \cdot\,.1 & & -1 & \cdot\,.-1
\end{bmatrix}
$$

and the error matrix is given by

$$
\epsilon = (\hat{\epsilon}_{11}, \ldots, \hat{\epsilon}_{dr}) .
$$

The maximum likelihood estimator of β, originally derived by Khatri (1966), can be translated into this setting as follows:

$$
\hat{\beta} = U\,B'\,V^{-1}\,Y\,A'\,A^* , \tag{7.5}
$$

where:

$$
V = Y\,(I - A'\,A^*\,A)\,Y' \tag{7.6}
$$

$$
U = (B'\,V^{-1}\,B)^{-1} \tag{7.7}
$$

$$
A^* = (A\,A')^{-1}. \tag{7.8}
$$

The estimator of the concentration required to produce a 50% mortality for a given time point, is the value of the concentration that yields a value of $Y = \pi/4$ radians. The value of the log of this estimator is given by

$$X = (\pi/4 - g_1(T)) / g_2(T) \tag{7.9}$$

$$\text{where: } g_1(T) = \sum_{m=1}^{q} \hat{\beta}_{m1} T^{-(m-1)} \tag{7.10}$$

$$g_2(T) = \sum_{m=1}^{q} \hat{\beta}_{m2} T^{-(m-1)} \tag{7.11}$$

$\hat{\beta}_{mj}$ = the estimator of β_{mj}, the elements of β of (7.5). Consequently, the estimator of the LC50 is 10^X.

For the example we are considering, where $q = 2$, we could define our 3×2 design matrix by

$$B = \begin{bmatrix} 1 & T_1^{-1} \\ . & . \\ . & . \\ . & . \\ 1 & T_t^{-1} \end{bmatrix} = \begin{bmatrix} 1 & 0.500 \\ 1 & 0.333 \\ 1 & 0.250 \end{bmatrix}.$$

With only this degree of representation it can be shown that the estimated model becomes:

$$\hat{Y}_{ijk} = 1.193 - 1.286 T_i^{-1} + 0.782 X_j - 0.935 X_j T_i^{-1} \tag{7.12}$$
$$+ \hat{\rho}_{1k} + \hat{\rho}_{2k} T_i^{-1},$$

where

$$\hat{\rho}_{11} = -\hat{\rho}_{12} = 0.0302, \text{ and } \hat{\rho}_{21} = -\hat{\rho}_{22} = -0.0853.$$

7.2   Confidence Limits for LC50

In order to calculate the confidence interval for an LC50 we appeal to the following result.

Theorem 7.1. Assume the general multivariate growth curve model: $Y = B\beta A + \epsilon$, where Y is a $t \times rd$ response matrix, B is a $t \times q$ time matrix, β is a $q \times (r+1)$ parameter matrix, A is an $(r+1) \times rd$ design matrix and ϵ is a $t \times rd$ error matrix such that if $\epsilon = (\dot{\epsilon}_1, . . ., \dot{\epsilon}_{rd})$, then the components of ϵ are independent $N_t(0, \Sigma)$. Then the roots of the quadratic polynomial inside the following probability statement determine the confidence limits for $\log(LC50)$:

$$P[aX^2 + bX + c \leq 0] = 1 - \alpha, \tag{7.13}$$

where

$$a = [g_2(T)]^2 - g_3(T) \cdot f_\alpha w_{22}$$

$$b = 2[(g_1(T) - \pi/4)g_2(T) - g_3(T)f_\alpha w_{12}]$$

$$c = [g_1(T) - \pi/4]^2 - f_\alpha g_3(T)w_{11} ,$$

and where $g_1(T)$ and $g_2(T)$ are defined in (7.10) and (7.11), respectively and

$$g_3(T) = b'Ub , \tag{7.14}$$

where

$$b' = (1, T^{-1}, \ldots, T^{-(q-1)}) \tag{7.15}$$

and

$$U = (B' V^{-1} B)^{-1}$$

$$V = Y(I - A'A^*A)Y'$$

$$A^* = (AA')^{-1}$$

w_{ij} = the components of $W = A^*$
$\quad + A^*AY'(V^{-1} - V^{-1}BUB'V^{-1})YA'A^*$

$f_\alpha = x_\alpha/(1-x_\alpha)$

x_α = the upper $(1-\alpha)100\%$ point of the largest root of a multivariate beta distribution

a and b are vectors of coefficients.

When the multivariate growth curve model is a polynomial of degree $(q-1)$ in inverse time then $a' = (1, X, 0, \ldots, 0)$.

The x_α values have been tabulated; see, for example, Srivastava and Carter (1983). An x_α value depends on υ_b (the dimension of the vector space spanned by the vectors b of interest), υ_a (the dimension of the vector space spanned by the vectors a of interest) and $n = rd - (r+1) - (t-q)$. When υ_a is 1, then the x_α value is related to the univariate F_α value by the equation:

$$\frac{x_\alpha}{1-x_\alpha} = \frac{\upsilon_b}{n-\upsilon_b+1} F_{\upsilon_b, n-\upsilon_b+1; \alpha} . \tag{7.16}$$

Note that for one fixed time point and a fixed X value we have $\upsilon_b = 1$ and $\upsilon_a = 1$.

If the roots of the quadratic polynomial in (7.13) are L1 and L2, and the coefficient of X^2 is positive, then the confidence interval for the log(LC50) is (L1, L2). Hence the confidence interval for LC50 is given by $(10^{L1}, 10^{L2})$. If the coefficient of X^2 is negative, then the confidence region for the LC50 is given by the values not in the region $(10^{L1}, 10^{L2})$; in general, such a region is of little practical use. The type of confidence interval given in (7.13) was first derived by Fieller (1940) and later generalized by Bennett (1959) for a multivariate response case and by Zerbe (1978) for the multiple regression situation.

For our example the values of the matrices U, V and W used in the analysis are given by

$$U = \begin{bmatrix} 0.3865 & -0.8770 \\ -0.8770 & 2.0007 \end{bmatrix},$$

$$V = \begin{bmatrix} 0.023 & -0.015 & -0.004 \\ -0.015 & 0.025 & 0.036 \\ -0.004 & 0.036 & 0.095 \end{bmatrix}$$

and

$$W = \begin{bmatrix} 0.1193 & 0.1400 & 0.0080 \\ 0.1400 & 0.4405 & 0.0178 \\ 0.0080 & 0.0178 & 0.0738 \end{bmatrix}.$$

Further calculations yield, for selected values of T, the point and interval estimates for LC50 given in Table 7.2.

Table 7.2
POINT ESTIMATES AND CONFIDENCE INTERVALS
FOR LC50 AT SPECIFIED TIME POINTS

Time	Estimate	Confidence Interval
2.0	5.60	(3.99 , 8.76)
2.5	1.83	(1.61 , 2.11)
3.0	1.02	(0.93 , 1.38)
3.5	0.84	(0.68 , 0.91)
4.0	0.70	(0.56 , 0.85)
6.0	0.49	(0.37 , 0.66)

The expected values of the mortality counts based on the model in (7.12) are given in Table 7.3. These values agree very closely with the observed values in Table 7.1. Note that the usual chi-squared goodness-of-fit tests should not be performed on the data as the mortality counts over time are not independent.

Table 7.3
EXPECTED MORTALITY COUNTS

| Block | Time (Days) | \multicolumn{7}{c}{Concentration ($\mu g/\ell$)} |
		0.10	0.20	0.30	0.50	1.00	2.00	2.50
1	2	0.49	0.98	1.33	1.84	2.62	3.49	3.79
	3	0.85	1.79	2.47	3.42	4.81	6.21	6.65
	4	1.06	2.28	3.13	4.30	5.94	7.48	7.93
2	2	0.60	1.13	1.50	2.03	2.84	3.73	4.03
	3	0.83	1.77	2.44	3.39	4.77	6.18	6.61
	4	0.96	2.13	2.97	4.13	5.77	7.33	7.79

Simultaneous $(1-\alpha)100\%$ confidence bands for the LC50 for all time points, can be derived from (7.13) but with a value of

$$f_\alpha = (q/(n-q+1))\ F_{q,(n-q+1);\alpha},$$

where q is the dimension of the vector space spanned by the vectors b. An example of a plot of such simultaneous confidence bands for LC50 value over time can be found in Carter and Hubert (1984).

7.3   A Test for the Degree of the Polynomial in Time

To test the null hypothesis that higher order terms in inverse time are zero, the test statistic is

$$L = \frac{|V|}{|V+H|} \cdot \frac{|B'V^{-1}B|}{|B'(V+H)^{-1}B|}, \tag{7.17}$$

where V is defined in (7.6) and $H = YA'A^*AY'$. (Notice that $V+H = YY'$.) Then the null hypothesis that terms of order q and higher are zero is rejected if $L > L_\alpha$, where L_α is the upper $\alpha 100\%$ point of Wilk's statistic with parameters $v_b = (t-q)$, $v_a = (r+1)$ and $v = r(d-1) - 1$, the degrees of freedom of error. For details of this test see Srivastava and Carter (1983), page 184.

The choice of a model in inverse time or direct time can be decided by the researcher and the data. In experiments where the subjects become acclimated to the treatments, there is little change in the response variable for large values of T; hence, a choice of inverse time seems appropriate.

For our example the test of the null hypothesis that the reduced model we have used is adequate versus the alternative hypothesis that the quadratic model is necessary, the test statistic in (7.17) must be calculated. The value of the H matrix is

$$H = \begin{bmatrix} 3.589 & 4.792 & 5.622 \\ 4.792 & 6.402 & 7.508 \\ 5.622 & 7.508 & 8.825 \end{bmatrix}$$

With the B and V matrices found earlier, the value of the test statistic (7.17) is L = 0.652, with parameters $v_b = t - q = 1$, $v_a = r + 1 = 3$ and $n = r(d-1) - 1 = 11$. Because $v_b = 1$ we can use the result that under the null hypothesis $F = (v/v_a)(1-L)/L$, where F has a Fisher F distribution with degrees of freedom v_a and v. In this case, F = 1.95. The upper 5% critical value is given by $F_{3,11;0.05} = 3.59$. Therefore, we do not reject the null hypothesis and claim that the quadratic term is not significant. The full model in inverse time is not necessary, and the reduced model is sufficient.

7.4 Concluding Remarks

In this chapter it has been demonstrated that polynomial growth curve models can be effectively employed to analyze multivariate quantal bioassays over time. Such a method assumes that a smooth curve approximates the relationship between the transformed quantal values and the two explanatory variables, log concentration and time. Moreover, this method performs well for interpolation problems, and is therefore ideal for estimating LC50 values. The values of the concentration should be chosen such that 0% and 100% mortality counts do not dominate the data. Time points should be chosen such that the correlation between responses at successive time points is not too close to one.

The general method of fitting growth curves to experiments with many time points has been thoroughly discussed by Rao (1965); also see Grizzle and Allen (1969). They conclude that the response variable can be modelled by orthogonal polynomials with linear, quadratic and higher order terms but that it is often the practice that only a few of the higher order terms are used in the analysis.

In the example considered in this chapter, the LC50 values were calculated with the block effects set to zero. The replicates (blocks) represented two consecutive weeks when the experiment was performed; this is consistent with the observation that the estimates of these effects were small. Testing for block effects can be done by applying standard MANOVA techniques.

An analysis of indirect quantitative assays with correlated data has been given by Box and Hay (1953). It should be noted that this method assumes a repeated measures model; that is, the correlations between observations on the same experimental unit are assumed to be equal. Elashoff (1981) studied the effect of heterogeneity of variances and correlations on this analysis. Vølund (1980) treated the case when these correlations are arbitrary, extending the work of Rao (1954). Multivariate indirect quantal assays have also been studied by Kolakowski and Bock (1981) and Kooijman (1981). Kolakowski and Bock (1981) assumed a latent structural model that assumed independence among the responses over time. Kooijman (1981) assumed that the mortality counts follow a multinomial distribution with the probability of death a function of time and concentration. Such a model does not differentiate between sampling units and experimental units. For example, if concentrations of copper are administered to tanks of 20 fish, the model assumes that there is no error variation among tanks of fish. Some work on including this extra variation in the model for experiments involving litters of subjects has been given in Segreti and Munson (1981).

The methods presented in this chapter should be extendable to nonlinear growth curve models and to the analysis of incomplete data in which measurements are not taken at every time point for each experimental unit. Also, the methods should be useful for generating confidence intervals for the relative potency in a multivariate quantitative parallel-line bioassay.

Appendices: Theorems

APPENDIX 1. FIELLER'S THEOREM

Suppose X_1 and X_2 are normal random variables with $E[X_i] = \mu_i$, $V[X_i] = c_{ii}\sigma^2$, for $i = 1, 2$ and $\text{Cov}[X_1, X_2] = c_{12}\sigma^2$. Let $R = X_1/X_2$ and $\phi = \mu_1/\mu_2$. Then the confidence limits for ϕ are given by

$$\frac{1}{1-g}\left\{ R - g\frac{c_{12}}{c_{22}} \pm \frac{ts}{X_2}\sqrt{c_{11}(1-g) - 2c_{12}R + c_{22}R^2 + g\,c_{12}^2/c_{22}} \right\}$$

where $g = \dfrac{t^2 s^2 c_{22}}{X_2^2}$, s is an estimator of σ, and t is the appropriate percentile point for a Student's t distribution with degrees of freedom depending on the estimator s.

Proof

Let $L = X_1 - rX_2$. Since $E[L] = \mu_1 - r\mu_2$, $V[L] = \sigma_1^2 + r^2\sigma_2^2 - 2r\sigma_{12}$ where $\sigma_i^2 = V[X_i] = c_{ii}\sigma^2$, $i = 1, 2$ and $\sigma_{12} = \text{Cov}[X_1, X_2] = c_{12}\sigma^2$, then for any r

$$P[R \leqslant r] = P\left[\frac{L - E[L]}{\sqrt{V[L]}} \leqslant \frac{-(\mu_1 - r\mu_2)}{\sqrt{\sigma_1^2 + r^2\sigma_2^2 - 2r\sigma_{12}}} \right].$$

If we choose $r = \mu_1/\mu_2 = \phi$ then L is a linear combination of normal variables with $E[L] = 0$ and $V[L] = \sigma_1^2 + \phi^2\sigma_2^2 + 2\phi\sigma_{12}$ so that $(L - E[L])/\sqrt{V[L]}$ is distributed as a $N(0, 1)$ variable. Also, if s^2 estimates σ^2 with ν degrees of freedom (d.f.), then it is well known that $\nu s^2/\sigma^2$ has a χ^2 distribution with ν d.f. Since the Student's t distribution (with ν d.f.) is defined by $Z/\sqrt{\chi^2/\nu}$, when Z is $N(0, 1)$, then

$$\frac{X_1 - \phi X_2}{s\sqrt{c_{11} + \phi^2 c_{22} - 2\phi c_{12}}}$$

is a t variate with ν d.f. Let t_α be the appropriate percentile point such that $P[-t_\alpha \leqslant t \leqslant t_\alpha] = 1 - \alpha$.

Now from $t_\alpha = (X_1 - \phi X_2)/s\sqrt{c_{11} + \phi^2 c_{22} - 2\phi c_{12}}$ it follows that

$$\phi^2(t_\alpha^2 s^2 c_{22} - X_2^2) - 2\phi(t_\alpha^2 s^2 c_{12} - X_1 X_2) + t_\alpha^2 s^2 c_{11} - X_1^2 = 0.$$

Define

$$g_{ii} = \frac{t_\alpha^2 s^2 c_{ii}}{X_i^2}, i = 1, 2; g_{12} = \frac{t_\alpha^2 s^2 c_{12}}{X_1 X_2}.$$

Then we obtain

$$\phi^2 X_2^2 (g_{22} - 1) - 2\phi X_1 X_2 (g_{12} - 1) + X_1^2(g_{11} - 1) = 0.$$

Thus the confidence limits for ϕ, denoted by ϕ_L, ϕ_U, are defined by

$$\frac{2X_1(g_{12} - 1) \pm \sqrt{(2X_1(g_{12} - 1))^2 - 4(g_{22} - 1) X_1^2 (g_{11} - 1)}}{2X_2(g_{22} - 1)}$$

When we replace X_1/X_2 by R we finally obtain after a little algebra that

$$\frac{R - g_{22} \dfrac{c_{12}}{c_{22}} \pm \dfrac{t_\alpha s}{X_2} \sqrt{c_{11}(1 - g_{22}) - 2c_{12} R + c_{22} R^2 + g (c_{12}^2 / c_{22})}}{1 - g_{22}}$$

the desired result.

APPENDIX 2.   THE BLISS THEOREM

Suppose $R = \overline{X}_1/\overline{X}_2$, X_1 and X_2 are independently distributed normal random variables, and $\phi = \dfrac{\mu_1}{\mu_2} = \rho$ (relative potency in direct assays) then the confidence limits of ρ are defined by

$$\frac{R \pm \dfrac{ts}{\overline{X}_2} \sqrt{\dfrac{1}{n_1}(1 - g) + \dfrac{R^2}{n_2}}}{1 - g}$$

where $g = t^2 s^2 / n_2 (\overline{X}_2)^2$.

Proof

We apply Fieller's Theorem (Appendix 1) where here $\sigma_1^2 = V[\overline{X}_1] = \dfrac{\sigma^2}{n_1} = c_{11}\sigma^2$ so that $c_{11} = 1/n_1$; $\sigma_2^2 = V[\overline{X}_2] = \sigma^2/n_2 = c_{22}\sigma^2$, so that $c_{22} = 1/n_2$; and $\sigma_{12} = Cov[\overline{X}_1, \overline{X}_2] = c_{12}\sigma^2 = 0$, so that $c_{12} = 0$. Consequently $g_{12} = 0$ and $g_{ii} = \dfrac{t^2 s^2}{n_i \overline{X}_i^2}$, $i = 1, 2$ and $g = g_{22}$. Substituting we obtain the desired result.

Corollary 2.1   Special Case

If $g \simeq 0$ and $n = n_1 = n_2$ then the above confidence limits for the relative potency reduce simply to

$$R \pm t\, S_R$$

where

$$S_R = \frac{s}{\overline{X}_2} \sqrt{\frac{1 + R^2}{n}}.$$

APPENDIX 3.   *LIMITS FOR LOG RELATIVE POTENCY*

Suppose $Z_i = \log X_i$ ($i = 1, 2$) are normally distributed, equal variance $\hat{\sigma}$ and that the sampling is random (independent), then the confidence limits for the log relative potency are defined by

$$R^* \pm t\,\hat{\sigma} \sqrt{\frac{1}{n_1} + \frac{1}{n_2}}$$

where

$$R^* = \overline{Z}_1 - \overline{Z}_2 = \frac{1}{n_1} \sum_{j=1}^{n_1} \log X_{j1} - \frac{1}{n_2} \sum_{j=1}^{n_2} \log X_{j2} \ .$$

Proof If we let Z_i be $N(\log \mu_i, \sigma^2)$, then for $i = 1, 2$,

$$\overline{Z}_i \sim N(\log \mu_i, \sigma^2/n_i), \ i = 1, 2,$$

and

$$\overline{Z}_1 - \overline{Z}_2 \sim N\left(\log(\mu_1/\mu_2), \sigma^2\left(\frac{1}{n_1} + \frac{1}{n_2}\right)\right).$$

Consequently

$$\frac{\overline{Z}_1 - \overline{Z}_2 - \log(\mu_1/\mu_2)}{\sigma \sqrt{\frac{1}{n_1} + \frac{1}{n_2}}} \sim t(\nu).$$

Let $R^* = \overline{Z}_1 - \overline{Z}_2$, and $\rho = \mu_1/\mu_2$. Then there exists a t value, t_α, say such that

$$P\left[-t_\alpha \leq \frac{R^* - \log \rho}{\hat{\sigma} \sqrt{\frac{1}{n_1} + \frac{1}{n_2}}} \leq t_\alpha \right] = 1 - \alpha.$$

By isolating $\log \rho$ within the inequality signs, it follows that the confidence limits for $\log \rho$ are given by

$$R^* \pm t\,\hat{\sigma} \sqrt{\frac{1}{n_1} + \frac{1}{n_2}} \ .$$

APPENDIX 4. $V[\hat{b}]$

$V[\hat{b}] = \sigma^2/D.$

Proof

Using the notation in chapter 3 we have

$$V[\hat{b}] = V\left[\frac{\Sigma x_s y_s + \Sigma x_t y_t}{\Sigma x_s^2 + \Sigma x^2}\right]$$

$$= \frac{1}{D^2} V[\Sigma(X_s - \overline{X}_s)Y_s + \Sigma(X_t - \overline{X}_t)Y_t]$$

$$= \frac{1}{D^2} \{\Sigma(X_s - \overline{X}_s)^2 V[Y_s] + \Sigma(X_t - \overline{X}_t)^2 V[Y_t]\}$$

$$= \frac{1}{D^2} \{\Sigma x_s^2 \sigma^2 + \Sigma x_t^2 \sigma^2\}$$

$$= \sigma^2/D.$$

APPENDIX 5. $E[M]$ *and* $V[M]$

The approximate values of the expectation and variance of $M = \log \hat{\rho}$, where ρ is the relative potency, are respectively

$$E[M] \doteq \log\rho\left\{1 + \frac{\sigma^2}{Db^2}\left(1 + \frac{\overline{X}_t - \overline{X}_s}{a_t - a_s}\right)\right\}$$

$$V[M] \doteq \frac{\sigma^2}{b^2}\left\{\frac{1}{n_t} + \frac{1}{n_s} + \frac{1}{D}(\overline{X}_t - \overline{X}_s + \log\rho)^2\right\}.$$

Proof

We use the notation of chapter 3 on parallel line assays, and consider

$$M = \log\hat{\rho} = \frac{\overline{Y}_t - \overline{Y}_s - \hat{b}(\overline{X}_t - \overline{X}_s)}{\hat{b}} = \frac{Z_1}{Z_2}, \text{ say.}$$

Notice that M is the ratio of random variables, i.e., a nonlinear function of random variables. Provided the samples are large, one avenue of approach is to apply the so-called 'propagation of error formula', which is defined as follows.

If $F(Z_1, Z_2)$ is a nonlinear function of two variables Z_1, Z_2 then for arbitrary values (μ_1, μ_2) of (Z_1, Z_2) we have

$$F(Z_1, Z_2) = F(\mu_1, \mu_2) + (Z_1 - \mu_1)F_1 + (Z_2 - \mu_2)F_2$$

$$+ \frac{1}{2}\{(Z_1 - \mu_1)^2 F_{11} + (Z_2 - \mu_2)^2 F_{22} - 2(Z_1 - \mu_1)(Z_2 - \mu_2)F_{12}\}$$

$$+ \text{ higher order terms in } Z_1 \text{ and } Z_2,$$

where

$$F_1 = \frac{\partial F(Z_1, Z_2)}{\partial Z_1} \text{ evaluated at } Z_1 = \mu_1, Z_2 = \mu_2$$

$$F_2 = \frac{\partial F(Z_1, Z_2)}{\partial Z_2} \text{ evaluated at } Z_1 = \mu_1, Z_2 = \mu_2.$$

Now when $F(Z_1, Z_2) = Z_1/Z_2$ then $F_1 = 1/\mu_2$, $F_2 = -\mu_1/\mu_2^2$, $F_{11} = 0$, $F_{22} = 2\mu_1/\mu_2^3$, $F_{12} = -1/\mu_2^2$. Disregarding the higher order terms we obtain

$$\frac{Z_1}{Z_2} = \frac{\mu_1}{\mu_2} + (Z_1 - \mu_1)\frac{1}{\mu_2} - (Z_2 - \mu_2)\frac{\mu_1}{\mu_2^2}$$

$$+ \frac{1}{2}\left\{(Z_2 - \mu_2)^2 \frac{2\mu_1}{\mu_2^3} - \frac{2(Z_1 - \mu_1)(Z_2 - \mu_2)}{\mu_2^2}\right\}$$

$$= \frac{\mu_1}{\mu_2}\left\{1 + \frac{(Z_1 - \mu_1)}{\mu_1} - \frac{(Z_2 - \mu_2)}{\mu_2} + \frac{(Z_2 - \mu_2)^2}{\mu_2^2} - \frac{(Z_1 - \mu_1)(Z_2 - \mu_2)}{\mu_1\mu_2}\right\}.$$

If we let $\mu_i = E[Z_i]$ (i = 1,2), then

$$E\left[\frac{Z_1}{Z_2}\right] = \frac{\mu_1}{\mu_2}\left\{1 + \frac{\sigma_2^2}{\mu_2^2} - \frac{\sigma_{12}}{\mu_1\mu_2}\right\}.$$

When

$$Z_1 = \overline{Y}_t - \overline{Y}_s - \hat{b}(\overline{X}_t - \overline{X}_s) \text{ and } Z_2 = \hat{b},$$

then

$$\mu_1 = E[Z_1] = E[(\overline{Y}_t - \hat{b}\,\overline{X}_t) - (\overline{Y}_s - \hat{b}\,\overline{X}_s)] = a_t - a_s$$

$$\mu_2 = E[Z_2] = E[\hat{b}] = b$$

$$\frac{\mu_1}{\mu_2} = \frac{a_t - a_s}{b} = \log \rho$$

$$\sigma_2^2 = V[Z_2] = V[\hat{b}] = \sigma^2/D$$

$$\sigma_1^2 = V[Z_1] = V[\overline{Y}_t - \overline{Y}_s] + (\overline{X}_t - \overline{X}_s)^2\, V[\hat{b}]$$

$$- 2\, \text{Cov}[(\overline{Y}_t - \overline{Y}_s), \hat{b}(\overline{X}_t - \overline{X}_s)]$$

$$= \frac{\sigma^2}{n_t} + \frac{\sigma^2}{n_s} + (\overline{X}_t - \overline{X}_s)^2 \frac{\sigma^2}{D}.$$

(Note, the covariance in σ_1^2 is zero.)

$$\sigma_{12} = \text{Cov}[Z_1, Z_2]$$
$$= \text{Cov}[(\overline{Y}_t - \overline{Y}_s) - \hat{b}(\overline{X}_t - \overline{X}_s), \hat{b}]$$
$$= -(\overline{X}_t - \overline{X}_s) \text{Cov}[\hat{b}, \hat{b}]$$
$$= -(\overline{X}_t - \overline{X}_s) V[\hat{b}]$$
$$= -(\overline{X}_t - \overline{X}_s) \cdot \frac{\sigma^2}{D}.$$

Hence

$$E\left[\frac{Z_1}{Z_2}\right] = \log\rho\left\{1 + \frac{\sigma^2}{Db^2} + \frac{(\overline{X}_t - \overline{X}_s)\sigma^2}{(a_t - a_s)bD}\right\}$$

or

$$E[M] = \log\rho\left\{1 + \frac{\sigma^2}{Db^2}\left(1 + \frac{\overline{X}_t - \overline{X}_s}{a_t - a_s}\right)\right\}.$$

For the variance of the estimator of the log relative potency, we disregard second order and higher order terms in the propagation of error formula to obtain the approximation

$$V[F(Z_1, Z_2)] = V[Z_1/Z_2] = V[M]$$
$$= \sigma_1^2 F_1^2 + \sigma_2^2 F_2^2 + 2\sigma_{12} F_1 F_2$$
$$= \sigma_1^2\left(\frac{1}{\mu_2}\right)^2 + \sigma_2^2\left(-\frac{\mu_1}{\mu_2^2}\right)^2 - 2\sigma_{12}\left(\frac{\mu_1}{\mu_2^3}\right)$$
$$= \frac{\sigma^2}{\mu_2^2}\left\{\frac{1}{n_t} + \frac{1}{n_s} + \frac{1}{D}\left[(\overline{X}_t - \overline{X}_s)^2 + \frac{\mu_1^2}{\mu_2^2} + \frac{2(\overline{X}_t - \overline{X}_s)\mu_1}{\mu_2}\right]\right\}.$$

Thus

$$V[M] = \frac{\sigma^2}{b^2}\left\{\frac{1}{n_t} + \frac{1}{n_s} + \frac{1}{D}(\overline{X}_t - \overline{X}_s + \log\rho)^2\right\}.$$

Corollary 5.1   An alternative form for $V[M]$

Since $\mu_1/\mu_2 = (a_t - a_s)/b = \log\rho$

then $$V[M] = V[Z_1/Z_2] = \left(\frac{\mu_1}{\mu_2}\right)^2\left\{\frac{\sigma_1^2}{\mu_1^2} + \frac{\sigma_2^2}{\mu_2^2} - \frac{2\sigma_{12}}{\mu_1\mu_2}\right\}$$
$$= \sigma^2(\log\rho)^2\left\{\frac{c_{11}}{\mu_1^2} + \frac{c_{22}}{\mu_2^2} - \frac{2c_{12}}{\mu_1\mu_2}\right\}$$

where $\sigma_1^2 = c_{11}\sigma^2$, $\sigma_2^2 = c_{22}\sigma^2$ and $\sigma_{12} = c_{12}\sigma^2$; thus

$$V[M] = \sigma^2 (\log \rho)^2 \left\{ \frac{\left(\frac{1}{n_t} + \frac{1}{n_s} \right) + \frac{1}{D}(\overline{X}_t - \overline{X}_s)^2}{(a_t - a_s)^2} + \frac{1}{Db^2} + \frac{2(\overline{X}_t - \overline{X}_s)}{Db(a_t - a_s)} \right\}$$

where a_t, a_s, b, σ and ρ can be estimated by $\hat{a}_t$, $\hat{a}_s$, $\hat{b}$, $\hat{\sigma}$ and $\hat{\rho}$ respectively.

Corollary 5.2   Variance of M for symmetric assays.

Whenever the assay is symmetrical and the dosage levels are the same for the standard and test preparations, then $\overline{X}_t = \overline{X}_s$. In this special case

$$E[\log \hat{\rho}] = \log \rho \left(1 + \frac{\sigma^2}{Db^2} \right)$$

$$V[\log \hat{\rho}] = \frac{\sigma^2}{b^2} \left(\frac{1}{n_t} + \frac{1}{n_s} + \frac{\log^2 \rho}{D} \right).$$

APPENDIX 6.   APPROXIMATE CONFIDENCE LIMITS FOR log ρ

Using the notation of chapter 3, on parallel assays, the approximate confidence limits for $\log \rho$ are defined by

$$\frac{1}{1 - g} \left\{ M + (\overline{X}_t - \overline{X}_s)g \pm \frac{t\hat{\sigma}}{\hat{b}} \sqrt{n^*(1 - g) + \frac{1}{D}[M + (\overline{X}_t - \overline{X}_s)]^2} \right\}$$

where

$$M = \log \hat{\rho}, \quad n^* = \frac{1}{n_t} + \frac{1}{n_s}, \quad g = \frac{t^2 \hat{\sigma}^2}{\hat{b}^2 D}.$$

Proof

We apply Fieller's Theorem (Appendix 1) to

$$\phi = \log \rho = \frac{a_t - a_s}{b}, \quad \text{which is estimated by}$$

$$M = \log \hat{\rho} = \frac{\hat{a}_t - \hat{a}_s}{\hat{b}} = \frac{(\overline{Y}_t - \overline{Y}_s) - \hat{b}(\overline{X}_t - \overline{X}_s)}{\hat{b}}$$

Now, in this case

$$\sigma_1^2 = c_{11}\sigma^2 = V[\overline{Y}_t - \overline{Y}_s - \hat{b}(\overline{X}_t - \overline{X}_s)] = \frac{\sigma^2}{n_t} + \frac{\sigma^2}{n_s} + (\overline{X}_t - \overline{X}_s)^2 \frac{\sigma^2}{D}$$

$$\sigma_2^2 = c_{22}\sigma^2 = V[\hat{b}] = \sigma^2/D$$

$$\sigma_{12} = c_{12}\sigma^2 = \text{Cov}[\hat{a}_t - \hat{a}_s, \hat{b}] = -\frac{(\overline{X}_t - \overline{X}_s)\sigma^2}{D},$$

so that

$$c_{11} = \frac{1}{n_t} + \frac{1}{n_s} + \frac{1}{D}(\overline{X}_t - \overline{X}_s)^2$$

$$c_{22} = 1/D$$

$$c_{12} = -(\overline{X}_t - \overline{X}_s)/D$$

$$g_{11} = \frac{t^2 \, \hat{\sigma}^2}{\hat{a}_t - \hat{a}_s}\left\{ \frac{1}{n_t} + \frac{1}{n_s} + \frac{1}{D}(\overline{X}_t - \overline{X}_s)^2 \right\}$$

$$g_{22} = \frac{t^2 \, \hat{\sigma}^2}{D\hat{b}^2}$$

$$g_{12} = \frac{-t^2 \, \hat{\sigma}^2 (\overline{X}_t - \overline{X}_s)}{D \, \hat{b} \, (\hat{a}_t - \hat{a}_s)}.$$

Therefore the approximate confidence limits for log ρ are given by

$$\frac{1}{1 - \dfrac{t^2 \, \hat{\sigma}^2}{\hat{b}^2 \, D}}\left\{ M - \frac{t^2 \, \hat{\sigma}^2}{D \, \hat{b}^2}\left(-\frac{\overline{X}_t - \overline{X}_s}{D} \right) D \pm \frac{t \, \hat{\sigma}}{\hat{b}}\sqrt{(*)} \right\},$$

where

$$(*) = \frac{1}{n_t} + \frac{1}{n_s} + \frac{1}{D}(\overline{X}_t - \overline{X}_s)^2 +$$

$$\frac{2M}{D}(\overline{X}_t - \overline{X}_s) + \frac{M^2}{D} - \frac{t^2 \, \hat{\sigma}^2}{\hat{b}^2 \, D}\left\{ \frac{1}{n_t} + \frac{1}{n_s} \right\}.$$

If we now let $n^* = \dfrac{1}{n_t} + \dfrac{1}{n_s}$ and $g = \dfrac{t^2 \, \hat{\sigma}^2}{\hat{b}^2 \, D}$ the desired result follows.

Note: Most assays are designed so that the dosage levels are the same for each preparation. In this case $\overline{X}_t = \overline{X}_s$. In this special case, the above confidence limits for log ρ reduce simply to

$$\frac{M \pm tS^*}{1 - g},$$

where

$$S^* = \frac{\hat{\sigma}}{\hat{b}}\sqrt{\left(\frac{1}{n_t} + \frac{1}{n_s} \right)(1 - g) + \frac{M^2}{D}}.$$

APPENDIX 7.   PARTITIONING THE SS

Using the notation of section 3.14 we have

Theorem

(a) $SS(T) = SS(1) + SS(E)$

(b) $SS(1) = \sum_{j=2}^{5} SS(j)$.

Proof

(a) This is the basic partitioning in CRD and follows from the fact that

$$SS(T) = \sum_{u=s}^{t} \sum_{i=1}^{k_u} \sum_{j=1}^{n_{ui}} (Y_{uij} - \overline{Y})^2$$

$$= \sum_{u} \sum_{i} \sum_{j} (Y_{uij} - \overline{Y}_{ui} + \overline{Y}_{ui} - \overline{Y})^2$$

$$= \sum_{u} \sum_{i} \sum_{j} (Y_{uij} - \overline{Y}_{ui})^2 + \sum_{u} \sum_{i} \sum_{j} (\overline{Y}_{ui} - \overline{Y})^2,$$

because the cross-product is

$$2 \sum_{u} \sum_{i} \sum_{j} (Y_{uij} - \overline{Y}_{ui}) (\overline{Y}_{ui} - \overline{Y})$$

$$= 2 \sum_{u} \sum_{i} (\overline{Y}_{ui} - \overline{Y}) \left(\sum_{j=1}^{n_{ui}} (Y_{uij} - \overline{Y}_{ui}) \right)$$

$= 0$, since the quantity within the parentheses is zero.

That is,

$$SS(T) = \sum_{u} \sum_{i} \sum_{j} (Y_{uij} - \overline{Y}_{ui})^2 + \sum_{u} \sum_{i} n_{ui} (\overline{Y}_{ui} - \overline{Y})^2$$

$$= SS(E) + SS(1).$$

(b) $SS(1) = \sum_{u} \sum_{i} n_{ui} (\overline{Y}_{ui} - \overline{Y})^2$

$$= \sum_{u} \sum_{i} n_{ui} [(\overline{Y}_{ui} - \overline{Y}_{u}) + (\overline{Y}_{u} - \overline{Y})]^2$$

$$= \sum_{u} \sum_{i} n_{ui} (\overline{Y}_{ui} - \overline{Y}_{u})^2 + \sum_{u} \sum_{i} n_{ui} (\overline{Y}_{u} - \overline{Y})^2,$$

since the cross-product is

$$\sum_u \sum_i n_{ui}(\overline{Y}_{ui} - \overline{Y}_u)(\overline{Y}_u - \overline{Y})$$

$$= \sum_u (\overline{Y}_u - \overline{Y}) \sum_i n_{ui}(\overline{Y}_{ui} - \overline{Y}_u)$$

$$= 0,$$

because the last factor is zero:

$$\sum_i n_{ui}\overline{Y}_{ui} - \overline{Y}_u \sum_i n_{ui}$$

$$= \sum_i n_{ui}\left(\sum_i Y_{uij}/n_{ui} \right) - \overline{Y}_u n_u$$

$$= \sum_i \sum_j Y_{uij} - \overline{Y}_u n_u$$

$$= \overline{Y}_u n_u - \overline{Y}_u n_u$$

$$= 0.$$

So

$$SS(1) = \sum_u \sum_i n_{ui}(\overline{Y}_{ui} - \overline{Y}_u)^2 + \sum_u n_u(\overline{Y}_u - \overline{Y})^2 \qquad (7.1)$$

where the last term of (7.1) is SS(4), the SS(preparations), and will be small if $\overline{Y}_s \simeq \overline{Y}_t$, and large otherwise. Now the first term of (7.1) can be written as

$$\sum_u \sum_i n_{ui} (\overline{Y}_{ui} - \hat{a}_u - \hat{b}_u X_{ui} + \hat{a}_u + \hat{b}_u X_{ui} - \overline{Y}_u)^2 \qquad (7.2)$$

$$= \sum_u \sum_i n_{ui}(\overline{Y}_{ui} - \hat{a}_u - \hat{b}_u X_{ui})^2 + \sum_u \sum_i n_{ui}$$
$$(\hat{a}_u + \hat{b}_u X_{ui} - \overline{Y}_u)^2$$

where the cross-product is zero, since $\hat{a}_u = \overline{Y}_u - \hat{b}_u\overline{X}_u$ then

$$\sum_u \sum_i n_{ui} (\overline{Y}_{ui} - \hat{a}_u - \hat{b}_u X_{ui} (\hat{a}_u + \hat{b}_u X_{ui} - \overline{Y}_u)$$

$$= \sum_u \sum_i n_{ui} [\overline{Y}_{ui} - \overline{y}_u - \hat{b}_u (X_{ui} - \overline{X}_u)] [\hat{b}_u (X_{ui} - \overline{X}_u)]$$

$$= \sum_u \sum_i n_{ui} (Y_{ui} - \overline{Y}_u)\hat{b}_u (X_{ui} - \overline{X}_u) - \sum_u \sum_i n_{ui} \hat{b}_u^2 (X_{ui} - \overline{X}_u)^2$$

$$= \sum_u \hat{b}_u \left\{ \sum_i n_{ui} (Y_{ui} - \overline{Y}_u) (X_{ui} - \overline{X}_u) - \hat{b}_u \sum_i n_{ui} (X_{ui} - \overline{X}_u)^2 \right\}$$

$$= \sum_u \hat{b}_u^2 \left\{ \sum_i n_{ui}(Y_{ui} - \overline{Y}_u)(X_{ui} - \overline{X}_u) \right.$$

$$\left. - \frac{\sum\limits_i n_{ui}(Y_{ui} - \overline{Y}_u)(X_{ui} - \overline{X}_u)}{\sum\limits_i n_{ui}(X_{ui} - \overline{X}_u)^2} \cdot \sum_i n_{ui}(X_{ui} - \overline{X}_u)^2 \right\}$$

$$= 0.$$

The first term of (7.2) is SS(2), which will be small if the $\overline{Y}_{ui}$ tend to be near their corresponding LS lines, and large otherwise. Since the lines used have slopes estimated from their respective slopes, we do not need parallelism.

The proof is complete if we show that the second term of (7.2) is SS(3) + SS(5). Now, this second term is

$$\sum_u \sum_i n_{ui}(\hat{a}_u + \hat{b}_u X_{ui} - \overline{Y}_u)^2$$

$$= \sum_u \sum_i n_{ui}(\overline{Y}_u - \hat{b}_u \overline{X}_u + \hat{b}_u X_{ui} - \overline{Y}_u)^2$$

$$= \sum_u \sum_i n_{ui} \hat{b}_u^2(X_{ui} - \overline{X}_u)^2$$

$$= \sum_u \hat{b}_u^2 \sum_i n_{ui}(X_{ui} - \overline{X}_u)^2$$

$$= \sum_u \hat{b}_u^2 w_u$$

$$= \sum_u w_u(\hat{b}_u - \hat{b} + \hat{b})^2$$

$$= \sum_u w_u(\hat{b}_u - \hat{b})^2 + \sum_u w_u \hat{b}^2, \qquad (7.3)$$

where the cross-product is zero since

$$\sum_u w_u(\hat{b}_u - \hat{b})\hat{b} = \hat{b}\left(\sum_u w_u \hat{b}_u - \hat{b} \sum_u w_u \right)$$

$$= \hat{b}\left(\hat{b} \sum_u w_u - \hat{b} \sum_u w_u \right)$$

$$= 0.$$

But (7.3) is SS(3) + SS(5), which completes the proof.

APPENDIX 8:   SPEARMAN–KÄRBER ESTIMATOR

Theorem

   (5.2) follows from (5.1).

Proof

   Since $p_1 = 0$, $p_k = 1$ and $d = X_{i+1} - X_i$, then

$$m = \frac{1}{2} \sum_{i=1}^{k-1} (p_{i+1} - p_i)\,(2X_i + d)$$

$$= \sum_{i=1}^{k-1} (p_{i+1} - p_i)\,X_i + \frac{d}{2} \sum_{i=1}^{k-1} (p_{i+1} - p_i)$$

$$= \sum_{i=1}^{k-1} p_{i+1}\,X_i - \sum_{j=1}^{k-2} p_{j+1}\,X_{j+1} + \frac{d}{2}\,(p_k - p_1)$$

$$= p_k\,X_k - \sum_{i=1}^{k-1} p_{i+1}\,(X_{i+1} - X_i) + \frac{d}{2}$$

$$= X_k + \frac{d}{2} - d \sum_{i=1}^{k-1} p_{i+1}$$

$$= X_k + d \left(\frac{1}{2} - \sum_{j=1}^{k} p_i \right).$$

Tables

Table 1

PROBIT OF A PERCENTAGE RESPONSE p, FOR p = 0 (.1) 100

	.0	.1	.2	.3	.4	.5	.6	.7	.8	.9
0		1.9095	2.1215	2.2518	2.3475	2.4238	2.4874	2.5423	2.5907	2.6339
1	2.6732	2.7092	2.7424	2.7733	2.8023	2.8295	2.8551	2.8795	2.9026	2.9247
2	2.9458	2.9660	2.9855	3.0042	3.0222	3.0396	3.0564	3.0727	3.0885	3.1039
3	3.1188	3.1333	3.1474	3.1612	3.1746	3.1877	3.2005	3.2130	3.2252	3.2372
4	3.2489	3.2604	3.2717	3.2827	3.2936	3.3042	3.3147	3.3250	3.3351	3.3450
5	3.3548	3.3644	3.3739	3.3832	3.3924	3.4015	3.4104	3.4192	3.4279	3.4364
6	3.4449	3.4532	3.4615	3.4696	3.4777	3.4856	3.4934	3.5012	3.5089	3.5164
7	3.5239	3.5313	3.5387	3.5459	3.5531	3.5602	3.5672	3.5742	3.5811	3.5879
8	3.5947	3.6014	3.6080	3.6146	3.6211	3.6276	3.6340	3.6403	3.6466	3.6528
9	3.6590	3.6652	3.6713	3.6773	3.6833	3.6892	3.6951	3.7010	3.7068	3.7125
10	3.7183	3.7240	3.7296	3.7352	3.7408	3.7463	3.7518	3.7572	3.7626	3.7680
11	3.7733	3.7786	3.7839	3.7891	3.7943	3.7995	3.8047	3.8098	3.8148	3.8199
12	3.8249	3.8299	3.8349	3.8398	3.8447	3.8496	3.8544	3.8592	3.8640	3.8688
13	3.8735	3.8783	3.8830	3.8876	3.8923	3.8969	3.9015	3.9061	3.9106	3.9151
14	3.9196	3.9241	3.9286	3.9330	3.9375	3.9419	3.9462	3.9506	3.9549	3.9593
15	3.9636	3.9679	3.9721	3.9764	3.9806	3.9848	3.9890	3.9932	3.9973	4.0015
16	4.0056	4.0097	4.0138	4.0178	4.0219	4.0259	4.0300	4.0340	4.0380	4.0419
17	4.0459	4.0499	4.0538	4.0577	4.0616	4.0655	4.0694	4.0732	4.0771	4.0809
18	4.0847	4.0885	4.0923	4.0961	4.0999	4.1036	4.1074	4.1111	4.1148	4.1185
19	4.1222	4.1259	4.1296	4.1332	4.1369	4.1405	4.1442	4.1478	4.1514	4.1550
20	4.1585	4.1621	4.1657	4.1692	4.1728	4.1763	4.1798	4.1833	4.1868	4.1903
21	4.1938	4.1972	4.2007	4.2041	4.2076	4.2110	4.2144	4.2178	4.2213	4.2246
22	4.2280	4.2314	4.2348	4.2381	4.2415	4.2448	4.2482	4.2515	4.2548	4.2581
23	4.2614	4.2647	4.2680	4.2713	4.2745	4.2778	4.2810	4.2843	4.2875	4.2908
24	4.2940	4.2972	4.3004	4.3036	4.3068	4.3100	4.3132	4.3163	4.3195	4.3227
25	4.3258	4.3290	4.3321	4.3352	4.3384	4.3415	4.3446	4.3477	4.3508	4.3539
26	4.3570	4.3601	4.3631	4.3662	4.3693	4.3723	4.3754	4.3784	4.3815	4.3845
27	4.3875	4.3906	4.3936	4.3966	4.3996	4.4026	4.4056	4.4086	4.4116	4.4145
28	4.4175	4.4205	4.4235	4.4264	4.4294	4.4323	4.4353	4.4382	4.4411	4.4441
29	4.4470	4.4499	4.4528	4.4557	4.4587	4.4616	4.4645	4.4673	4.4702	4.4731
30	4.4760	4.4789	4.4817	4.4846	4.4875	4.4903	4.4932	4.4960	4.4989	4.5017
31	4.5046	4.5074	4.5102	4.5131	4.5159	4.5187	4.5215	4.5243	4.5271	4.5299
32	4.5327	4.5355	4.5383	4.5411	4.5439	4.5467	4.5494	4.5522	4.5550	4.5578
33	4.5605	4.5633	4.5660	4.5699	4.5715	4.5743	4.5770	4.5798	4.5825	4.5852
34	4.5880	4.5907	4.5934	4.5962	4.5989	4.6016	4.6043	4.6070	4.6097	4.6124
35	4.6151	4.6178	4.6205	4.6232	4.6259	4.6286	4.6313	4.6340	4.6366	4.6393

Table 1 (Continued)

	.0	.1	.2	.3	.4	.5	.6	.7	.8	.9
36	4.6420	4.6447	4.6473	4.6500	4.6527	4.6553	4.6580	4.6606	4.6633	4.6659
37	4.6686	4.6712	4.6739	4.6765	4.6792	4.6818	4.6844	4.6871	4.6897	4.6923
38	4.6950	4.6976	4.7002	4.7028	4.7054	4.7081	4.7107	4.7133	4.7159	4.7185
39	4.7211	4.7237	4.7263	4.7289	4.7315	4.7341	4.7367	4.7393	4.7414	4.7445
40	4.7471	4.7497	4.7522	4.7548	4.7574	4.7600	4.7626	4.7651	4.7677	4.7703
41	4.7729	4.7754	4.7780	4.7806	4.7831	4.7857	4.7883	4.7908	4.7934	4.7959
42	4.7985	4.8010	4.8036	4.8061	4.8087	4.8112	4.8138	4.8163	4.8189	4.8214
43	4.8240	4.8276	4.8291	4.8316	4.8341	4.8367	4.8392	4.8417	4.8443	4.8468
44	4.8493	4.8519	4.8544	4.8569	4.8595	4.8620	4.8645	4.8670	4.8696	4.8721
45	4.8746	4.8771	4.8797	4.8822	4.8847	4.8872	4.8897	4.8923	4.8948	4.8973
46	4.8993	4.9023	4.9048	4.9073	4.9099	4.9124	4.9149	4.9174	4.9199	4.9224
47	4.9249	4.9274	4.9299	4.9324	4.9349	4.9375	4.9400	4.9425	4.9450	4.9475
48	4.9500	4.9525	4.9550	4.9575	4.9600	4.9625	4.9650	4.9675	4.9700	4.9725
49	4.9750	4.9775	4.9800	4.9825	4.9850	4.9875	4.9900	4.9925	4.9950	4.9975
50	5.0000	5.0025	5.0050	5.0075	5.0100	5.0125	5.0150	5.0175	5.0200	5.0225
51	5.0250	5.0275	5.0300	5.0325	5.0350	5.0375	5.0400	5.0425	5.0450	5.0475
52	5.0500	5.0525	5.0550	5.0575	5.0600	5.0625	5.0651	5.0676	5.0701	5.0726
53	5.0751	5.0776	5.0801	5.0826	5.0851	5.0876	5.0901	5.0927	5.0952	5.0977
54	5.1002	5.1027	5.1052	5.1077	5.1103	5.1128	5.1153	5.1178	5.1203	5.1229
55	5.1254	5.1279	5.1304	5.1330	5.1355	5.1380	5.1405	5.1431	5.1456	5.1481
56	5.1507	5.1532	5.1557	5.1583	5.1608	5.1633	5.1659	5.1684	5.1709	5.1735
57	5.1760	5.1786	5.1811	5.1837	5.1862	5.1888	5.1913	5.1939	5.1964	5.1990
58	5.2015	5.2041	5.2066	5.2092	5.2118	5.2143	5.2169	5.2194	5.2220	4.2246
59	5.2271	5.2297	5.2323	5.2349	5.2374	5.2400	5.2426	5.2452	5.2478	5.2503
60	5.2529	5.2555	5.2581	5.2607	5.2633	5.2659	5.2685	5.2711	5.2737	5.2763
61	5.2789	5.2815	5.2841	5.2867	5.2893	5.2919	5.2946	5.2972	5.2998	5.3024
62	5.3050	5.3077	5.3103	5.3129	5.3156	5.3182	5.3208	5.3235	5.3261	5.3288
63	5.3314	5.3341	5.3367	5.3394	5.3420	5.3447	5.3473	5.3500	5.3527	5.3553
64	5.3580	5.3607	5.3634	5.3660	5.3687	5.3714	5.3741	5.3768	5.3795	5.3822
65	5.3849	5.3876	5.3903	5.3930	5.3957	5.3984	5.4011	5.4038	5.4066	5.4093
66	5.4120	5.4148	5.4175	5.4202	5.4230	5.4257	5.4285	5.4312	5.4340	5.4367
67	5.4395	5.4422	5.4450	5.4478	5.4506	5.4533	5.4561	5.4589	5.4617	5.4645
68	5.4673	5.4701	5.4729	5.4757	5.4785	5.4813	5.4841	5.4869	5.4898	5.4926
69	5.4954	5.4983	5.5011	5.5040	5.5068	5.5097	5.5125	5.5154	5.5183	5.5211
70	5.5240	5.5269	5.5298	5.5327	5.5355	5.5384	5.5413	5.5443	5.5472	5.5501
71	5.5530	5.5559	5.5589	5.5618	5.5647	5.5677	5.5706	5.5736	5.5765	5.5795
72	5.5825	5.5855	5.5884	5.5914	5.5944	5.5974	5.6004	5.6034	5.6064	5.6094
73	5.6125	5.6155	5.6185	5.6216	5.6246	5.6277	5.6307	5.6338	5.6369	5.6399
74	5.6430	5.6461	5.6492	5.6523	5.6554	5.6585	5.6616	5.6648	5.6679	5.6710
75	5.6742	5.6773	5.6805	5.6837	5.6868	5.6900	5.6932	5.6964	5.6996	5.7028

Table 1 (Continued)

	.0	.1	.2	.3	.4	.5	.6	.7	.8	.9
76	5.7060	5.7092	5.7125	5.7157	5.7190	5.7222	5.7255	5.7287	5.7320	5.7353
77	5.7386	5.7419	5.7452	5.7485	5.7518	5.7552	5.7585	5.7619	5.7652	5.7686
78	5.7720	5.7754	5.7787	5.7822	5.7856	5.7890	5.7924	5.7959	5.7993	5.8028
79	5.8062	5.8097	5.8132	5.8167	5.8202	5.8237	5.8272	5.8308	5.8343	5.8379
80	5.8415	5.8450	5.8486	5.8522	5.8558	5.8595	5.8631	5.8668	5.8704	5.8741
81	5.8778	5.8815	5.8852	5.8889	5.8926	5.8964	5.9001	5.9039	5.9077	5.9115
82	5.9153	5.9191	5.9229	5.9268	5.9306	5.9345	5.9384	5.9423	5.9462	5.9501
83	5.9541	5.9581	5.9620	5.9660	5.9700	5.9741	5.9781	5.9822	5.9862	5.9903
84	5.9944	5.9985	6.0027	6.0068	6.0110	6.0152	6.0194	6.0236	6.0279	6.0321
85	6.0364	6.0407	6.0451	6.0494	6.0538	6.0581	6.0625	6.0670	6.0714	6.0759
86	6.0804	6.0849	6.0894	6.0939	6.0985	6.1031	6.1077	6.1124	6.1170	6.1217
87	6.1265	6.1312	6.1360	6.1408	6.1456	6.1504	6.1553	6.1602	6.1651	6.1701
88	6.1751	6.1801	6.1852	6.1902	6.1953	6.2005	6.2057	6.2109	6.2161	6.2214
89	6.2267	6.2320	6.2374	6.2428	6.2482	6.2537	6.2592	6.2648	6.2704	6.2760
90	6.2817	6.2875	6.2932	6.2990	6.3049	6.3108	6.3167	6.3227	6.3287	6.3348
91	6.3410	6.3472	6.3534	6.3597	6.3660	6.3724	6.3789	6.3854	6.3920	6.3986
92	6.4053	6.4121	6.4189	6.4258	6.4328	6.4398	6.4469	6.4541	6.4613	6.4687
93	6.4761	6.4836	6.4911	6.4988	6.5066	6.5144	6.5223	6.5304	6.5385	6.5468
94	6.5551	6.5636	6.5721	6.5808	6.5896	6.5985	6.6076	6.6168	6.6261	6.6356
95	6.6452	6.6550	6.6649	6.6750	6.6853	6.6958	6.7064	6.7163	6.7283	6.7396
96	6.7511	6.7628	6.7748	6.7870	6.7995	6.8123	6.8254	6.8388	6.8526	6.8667
97	6.8812	6.8961	6.9115	6.9273	6.9436	6.9604	6.9778	6.9958	7.0145	7.0340
98	7.0542	7.0753	7.0974	7.1205	7.1449	7.1705	7.1977	7.2267	7.2576	7.2908
99	7.3268	7.3661	7.4093	7.4577	7.5126	7.5762	7.6525	7.7482	7.8785	8.0905

Table 2

THE WEIGHTING COEFFICIENT FOR A PROBIT(p) = Y = 1(.01)9

	.00	.01	.02	.03	.04	.05	.06	.07	.08	.09
1.0	.0006	.0006	.0006	.0006	.0007	.0007	.0007	.0007	.0008	.0008
1.1	.0008	.0009	.0009	.0009	.0009	.0010	.0010	.0011	.0011	.0011
1.2	.0012	.0012	.0013	.0013	.0014	.0014	.0015	.0015	.0016	.0016
1.3	.0017	.0017	.0018	.0019	.0019	.0020	.0021	.0021	.0022	.0023
1.4	.0024	.0024	.0025	.0026	.0027	.0028	.0029	.0030	.0031	.0032
1.5	.0033	.0034	.0035	.0036	.0037	.0038	.0040	.0041	.0042	.0044
1.6	.0045	.0047	.0048	.0050	.0051	.0053	.0054	.0056	.0058	.0060
1.7	.0061	.0063	.0065	.0067	.0069	.0071	.0074	.0076	.0078	.0080
1.8	.0083	.0085	.0088	.0090	.0093	.0096	.0099	.0101	.0104	.0107
1.9	.0110	.0114	.0117	.0120	.0124	.0127	.0131	.0134	.0138	.0142
2.0	.0146	.0150	.0154	.0158	.0162	.0167	.0171	.0176	.0181	.0185
2.1	.0190	.0195	.0200	.0206	.0211	.0217	.0222	.0228	.0234	.0240
2.2	.0246	.0252	.0258	.0265	.0272	.0278	.0285	.0292	.0299	.0307
2.3	.0314	.0322	.0330	.0338	.0346	.0354	.0362	.0371	.0380	.0389
2.4	.0398	.0407	.0416	.0426	.0436	.0446	.0456	.0466	.0476	.0487
2.5	.0498	.0509	.0520	.0532	.0543	.0555	.0567	.0579	.0591	.0604
2.6	.0617	.0630	.0643	.0656	.0670	.0684	.0698	.0712	.0727	.0741
2.7	.0756	.0771	.0787	.0802	.0818	.0834	.0851	.0867	.0884	.0901
2.8	.0918	.0935	.0953	.0971	.0989	.1007	.1026	.1045	.1064	.1083
2.9	.1103	.1122	.1142	.1163	.1183	.1204	.1225	.1246	.1268	.1289
3.0	.1311	.1333	.1356	.1378	.1401	.1424	.1448	.1471	.1495	.1519
3.1	.1544	.1568	.1593	.1618	.1643	.1669	.1694	.1720	.1746	.1773
3.2	.1799	.1826	.1853	.1881	.1908	.1936	.1964	.1992	.2020	.2049
3.3	.2077	.2106	.2135	.2165	.2194	.2224	.2254	.2284	.2314	.2345
3.4	.2375	.2406	.2437	.2468	.2500	.2531	.2563	.2594	.2626	.2658
3.5	.2691	.2723	.2756	.2788	.2821	.2854	.2887	.2920	.2953	.2986
3.6	.3020	.3053	.3087	.3121	.3155	.3188	.3222	.3256	.3291	.3325
3.7	.3359	.3393	.3428	.3462	.3496	.3531	.3565	.3600	.3634	.3669
3.8	.3703	.3738	.3772	.3807	.3841	.3876	.3910	.3944	.3979	.4013
3.9	.4047	.4082	.4116	.4150	.4184	.4218	.4252	.4285	.4319	.4353
4.0	.4386	.4420	.4453	.4486	.4519	.4552	.4585	.4617	.4650	.4682
4.1	.4714	.4746	.4778	.4810	.4841	.4873	.4904	.4935	.4965	.4996
4.2	.5026	.5056	.5086	.5116	.5145	.5174	.5203	.5232	.5260	.5288
4.3	.5316	.5343	.5371	.5398	.5425	.5451	.5477	.5503	.5529	.5554
4.4	.5579	.5603	.5628	.5652	.5675	.5699	.5722	.5744	.5766	.5788
4.5	.5810	.5831	.5852	.5872	.5893	.5912	.5932	.5951	.5969	.5987
4.6	.6005	.6023	.6040	.6056	.6072	.6088	.6104	.6119	.6133	.6147
4.7	.6161	.6174	.6187	.6199	.6211	.6223	.6234	.6245	.6255	.6265
4.8	.6274	.6283	.6292	.6300	.6307	.6314	.6321	.6327	.6333	.6338
4.9	.6343	.6347	.6351	.6355	.6358	.6360	.6362	.6364	.6365	.6366

Table 2 (Continued)

	.00	.01	.02	.03	.04	.05	.06	.07	.08	.09
5.0	.6366	.6366	.6365	.6364	.6362	.6360	.6358	.6355	.6351	.6347
5.1	.6343	.6338	.6333	.6327	.6321	.6314	.6307	.6300	.6292	.6283
5.2	.6274	.6265	.6255	.6245	.6234	.6223	.6211	.6199	.6187	.6174
5.3	.6161	.6147	.6133	.6119	.6104	.6088	.6072	.6056	.6040	.6023
5.4	.6005	.5987	.5969	.5951	.5932	.5912	.5893	.5872	.5852	.5381
5.5	.5810	.5788	.5766	.5744	.5722	.5699	.5675	.5652	.5628	.5603
5.6	.5579	.5554	.5529	.5503	.5477	.5451	.5425	.5398	.5371	.5343
5.7	.5316	.5288	.5260	.5232	.5203	.5174	.5145	.5116	.5086	.5056
5.8	.5026	.4996	.4965	.4935	.4904	.4873	.4841	.4810	.4778	.4746
5.9	.4714	.4682	.4650	.4617	.4585	.4552	.4519	.4486	.4453	.4420
6.0	.4386	.4353	.4319	.4285	.4252	.4218	.4184	.4150	.4116	.4082
6.1	.4047	.4013	.3979	.3944	.3910	.3876	.3841	.3807	.3772	.3738
6.2	.3703	.3669	.3634	.3600	.3565	.3531	.3496	.3462	.3428	.3393
6.3	.3359	.3325	.3291	.3256	.3222	.3188	.3155	.3121	.3087	.3053
6.4	.3020	.2986	.2953	.2920	.2887	.2854	.2821	.2788	.2756	.2723
6.5	.2691	.2658	.2626	.2594	.2563	.2531	.2500	.2468	.2437	.2406
6.6	.2375	.2345	.2314	.2284	.2254	.2224	.2194	.2165	.2135	.2106
6.7	.2077	.2049	.2020	.1992	.1964	.1936	.1908	.1881	.1853	.1826
6.8	.1799	.1773	.1746	.1720	.1694	.1669	.1643	.1618	.1593	.1568
6.9	.1544	.1519	.1495	.1471	.1448	.1424	.1401	.1378	.1356	.1333
7.0	.1311	.1289	.1268	.1246	.1225	.1204	.1183	.1163	.1142	.1122
7.1	.1103	.1083	.1064	.1045	.1026	.1007	.0989	.0971	.0953	.0935
7.2	.0918	.0901	.0884	.0867	.0851	.0834	.0818	.0802	.0787	.0771
7.3	.0756	.0741	.0727	.0712	.0698	.0684	.0670	.0656	.0643	.0630
7.4	.0617	.0604	.0591	.0579	.0567	.0555	.0543	.0532	.0520	.0509
7.5	.0498	.0487	.0476	.0466	.0456	.0446	.0436	.0426	.0416	.0407
7.6	.0398	.0389	.0380	.0371	.0362	.0354	.0346	.0338	.0330	.0322
7.7	.0314	.0307	.0299	.0292	.0285	.0278	.0272	.0265	.0258	.0252
7.8	.0246	.0240	.0234	.0228	.0222	.0217	.0211	.0206	.0200	.0195
7.9	.0190	.0185	.0181	.0176	.0171	.0167	.0162	.0158	.0154	.0150
8.0	.0146	.0142	.0138	.0134	.0131	.0127	.0124	.0120	.0117	.0114
8.1	.0110	.0107	.0104	.0101	.0099	.0096	.0093	.0090	.0088	.0085
8.2	.0083	.0080	.0078	.0076	.0074	.0071	.0069	.0067	.0065	.0063
8.3	.0061	.0060	.0058	.0056	.0054	.0053	.0051	.0050	.0048	.0047
8.4	.0045	.0044	.0042	.0041	.0040	.0038	.0037	.0036	.0035	.0034
8.5	.0033	.0032	.0031	.0030	.0029	.0028	.0027	.0026	.0025	.0024
8.6	.0024	.0023	.0022	.0021	.0021	.0020	.0019	.0019	.0018	.0017
8.7	.0017	.0016	.0016	.0015	.0015	.0014	.0014	.0013	.0013	.0012
8.8	.0012	.0011	.0011	.0011	.0010	.0010	.0009	.0009	.0009	.0009
8.9	.0008	.0008	.0008	.0007	.0007	.0007	.0007	.0006	.0006	.0006

Table 3
y, THE WORKING PROBITS FOR A GIVEN RESPONSE RATE, p%, AND EXPECTED PROBIT, Y

Y	p = 0	p = .01	p = .02	p = .03	p = .04	p = .05	p = .06	p = .07
.3	0.0955	–	–	–	–	–	–	–
.4	0.1915	–	–	–	–	–	–	–
.5	0.2872	–	–	–	–	–	–	–
.6	0.3828	–	–	–	–	–	–	–
.7	0.4782	–	–	–	–	–	–	–
.8	0.5734	–	–	–	–	–	–	–
.9	0.6684	–	–	–	–	–	–	–
1.0	0.7632	–	–	–	–	–	–	–
1.1	0.8578	–	–	–	–	–	–	–
1.2	0.9521	–	–	–	–	–	–	–
1.3	1.0462	–	–	–	–	–	–	–
1.4	1.1399	–	–	–	–	–	–	–
1.5	1.2334	–	–	–	–	–	–	–
1.6	1.3265	9.4420	–	–	–	–	–	–
1.7	1.4193	7.2246	–	–	–	–	–	–
1.8	1.5118	5.7062	9.9007	–	–	–	–	–
1.9	1.6038	4.6649	7.7260	–	–	–	–	–
2.0	1.6954	3.9518	6.2082	8.4646	–	–	–	–
2.1	1.7865	3.4665	5.1465	6.8264	8.5064	–	–	–
2.2	1.8772	3.1405	4.4039	5.6672	6.9306	8.1939	9.4573	–
2.3	1.9673	2.9269	3.8865	4.8461	5.8057	6.7653	7.7249	8.6845
2.4	2.0568	2.7930	3.5293	4.2655	5.0017	5.7379	6.4741	7.2103
2.5	2.1457	2.7162	3.2867	3.8573	4.4278	4.9983	5.5688	6.1393
2.6	2.2339	2.6805	3.1270	3.5736	4.0201	4.4666	4.9132	5.3597
2.7	2.3214	2.6744	3.0275	3.3805	3.7335	4.0865	4.4395	4.7926
2.8	2.4081	2.6900	2.9719	3.2538	3.5356	3.8175	4.0994	4.3813
2.9	2.4938	2.7212	2.9486	3.1759	3.4033	3.6306	3.8580	4.0853
3.0	2.5786	2.7638	2.9491	3.1343	3.3195	3.5047	3.6899	3.8751
3.1	2.6624	2.8148	2.9672	3.1196	3.2720	3.4244	3.5768	3.7292
3.2	2.7449	2.8716	2.9982	3.1249	3.2515	3.3782	3.5049	3.6315
3.3	2.8261	2.9325	3.0388	3.1451	3.2515	3.3578	3.4641	3.5704
3.4	2.9060	2.9961	3.0863	3.1764	3.2666	3.3567	3.4469	3.5370
3.5	2.9842	3.0614	3.1386	3.2158	3.2930	3.3702	3.4474	3.5247
3.6	3.0606	3.1274	3.1942	3.2610	3.3278	3.3946	3.4614	3.5282
3.7	3.1351	3.1935	3.2518	3.3102	3.3685	3.4269	3.4853	3.5436
3.8	3.2074	3.2589	3.3104	3.3619	3.4134	3.4649	3.5164	3.5679
3.9	3.2773	3.3232	3.3691	3.4150	3.4609	3.5068	3.5527	3.5986
4.0	3.3443	3.3856	3.4270	3.4683	3.5096	3.5510	3.5923	3.6336
4.1	3.4083	3.4458	3.4834	3.5210	3.5586	3.5962	3.6338	3.6713
4.2	3.4687	3.5032	3.5377	3.5722	3.6068	3.6413	3.6758	3.7103
4.3	3.5251	3.5571	3.5892	3.6212	3.6532	3.6852	3.7173	3.7493
4.4	3.5770	3.6070	3.6370	3.6670	3.6970	3.7270	3.7570	3.7870
4.5	3.6236	3.6520	3.6804	3.7088	3.7373	3.7657	3.7941	3.8225

Table 3 (Continued)

Y	p = 0	p = .01	p = .02	p = .03	p = .04	p = .05	p = .06	p = .07
4.6	3.6643	3.6915	3.7186	3.7458	3.7729	3.8001	3.8273	3.8544
4.7	3.6982	3.7244	3.7506	3.7768	3.8030	3.8293	3.8555	3.8817
4.8	3.7241	3.7496	3.7752	3.8008	3.8263	3.8519	3.8775	3.9031
4.9	3.7407	3.7659	3.7911	3.8163	3.8415	3.8667	3.8919	3.9171
5.0	3.7467	3.7718	3.7968	3.8219	3.8470	3.8720	3.8971	3.9222
5.1	3.7401	3.7653	3.7905	3.8156	3.8408	3.8660	3.8912	3.9164
5.2	3.7187	3.7443	3.7698	3.7954	3.8210	3.8465	3.8721	3.8977
5.3	3.6798	3.7061	3.7323	3.7585	3.7847	3.8109	3.8372	3.8634
5.4	3.6203	3.6474	3.6746	3.7017	3.7289	3.7560	3.7832	3.8103
5.5	3.5360	3.5644	3.5928	3.6212	3.6496	3.6780	3.7064	3.7348
5.6	3.4220	3.4521	3.4821	3.5121	3.5421	3.5721	3.6021	3.6321
5.7	3.2724	3.3044	3.3364	3.3684	3.4005	3.4325	3.4645	3.4965
5.8	3.0794	3.1139	3.1484	3.1829	3.2175	3.2520	3.2865	3.3210
5.9	2.8335	2.8711	2.9087	2.9463	2.9839	3.0214	3.0590	3.0966
6.0	2.5229	2.5643	2.6056	2.6469	2.6883	2.7296	2.7709	2.8122
6.1	2.1325	2.1784	2.2243	2.2702	2.3161	2.3620	2.4079	2.4538
6.2	1.6429	1.6944	1.7459	1.7974	1.8489	1.9004	1.9519	2.0034
6.3	1.0295	1.0878	1.1462	1.2046	1.2629	1.3213	1.3796	1.4380
6.4	0.2606	0.3273	0.3941	0.4609	0.5277	0.5945	0.6613	0.7281
6.5	–	–	–	–	–	–	–	–

Table 3 (Continued)

Y	p = .08	p = .09	p = .10	p = .11	p = .12	p = .13	p = .14	p = .15
2.1	–	–	–	–	–	–	–	–
2.2	–	–	–	–	–	–	–	–
2.3	9.6442	–	–	–	–	–	–	–
2.4	7.9466	8.6828	9.4190	–	–	–	–	–
2.5	6.7098	7.2803	7.8508	8.4213	8.9918	9.5623	–	–
2.6	5.8063	6.2528	6.6993	7.1459	7.5924	8.0389	8.4855	8.9320
2.7	5.1456	5.4986	5.8516	6.2046	6.5577	6.9107	7.2637	7.6167
2.8	4.6632	4.9451	5.2270	5.5089	5.7908	6.0727	6.3546	6.6365
2.9	4.3127	4.5401	4.7674	4.9948	5.2221	5.4495	5.6768	5.9042
3.0	4.0604	4.2456	4.4308	4.6160	4.8012	4.9864	5.1717	5.3569
3.1	3.8816	4.0340	4.1864	4.3388	4.4912	4.6436	4.7960	4.9484
3.2	3.7582	3.8849	4.0115	4.1382	4.2648	4.3915	4.5182	4.6448
3.3	3.6768	3.7831	3.8894	3.9957	4.1021	4.2084	4.3147	4.4211
3.4	3.6272	3.7173	3.8075	3.8977	3.9878	4.0780	4.1681	4.2583
3.5	3.6019	3.6791	3.7563	3.8335	3.9107	3.9879	4.0651	4.1423
3.6	3.5949	3.6617	3.7285	3.7953	3.8621	3.9289	3.9957	4.0625
3.7	3.6020	3.6603	3.7187	3.7770	3.8354	3.8937	3.9521	4.0104
3.8	3.6194	3.6709	3.7224	3.7739	3.8254	3.8769	3.9284	3.9799
3.9	3.6445	3.6904	3.7363	3.7822	3.8281	3.8740	3.9199	3.9658
4.0	3.6749	3.7163	3.7576	3.7989	3.8402	3.8816	3.9229	3.9642
4.1	3.7089	3.7465	3.7841	3.8217	3.8592	3.8968	3.9344	3.9720
4.2	3.7448	3.7794	3.8139	3.8484	3.8829	3.9174	3.9520	3.9865
4.3	3.7813	3.8133	3.8454	3.8774	3.9094	3.9414	3.9735	4.0055
4.4	3.8171	3.8471	3.8771	3.9071	3.9371	3.9671	3.9971	4.0271
4.5	3.8509	3.8793	3.9077	3.9361	3.9645	3.9929	4.0213	4.0497
4.6	3.8816	3.9087	3.9359	3.9630	3.9902	4.0173	4.0445	4.0716
4.7	3.9079	3.9341	3.9604	3.9866	4.0128	4.0390	4.0652	4.0915
4.8	3.9286	3.9542	3.9798	4.0054	4.0309	4.0565	4.0821	4.1076
4.9	3.9423	3.9675	3.9927	4.0178	4.0430	4.0682	4.0934	4.1186
5.0	3.9472	3.9723	3.9973	4.0224	4.0475	4.0725	4.0976	4.1227
5.1	3.9416	3.9668	3.9920	4.0172	4.0424	4.0676	4.0928	4.1179
5.2	3.9233	3.9488	3.9744	4.0000	4.0256	4.0511	4.0767	4.1023
5.3	3.8896	3.9158	3.9420	3.9683	3.9945	4.0207	4.0469	4.0731
5.4	3.83–5	3.8647	3.8918	3.9190	3.9461	3.9733	4.0004	4.0276
5.5	3.7632	3.7916	3.8200	3.8484	3.8768	3.9052	3.9336	3.9620
5.6	3.6621	3.6921	3.7221	3.7522	3.7822	3.8122	3.8422	3.8722
5.7	3.5286	3.5606	3.5926	3.6246	3.6567	3.6887	3.7207	3.7527
5.8	3.3555	3.3900	3.4246	3.4591	3.4936	3.5281	3.5626	3.5972
5.9	3.1342	3.1718	3.2094	3.2469	3.2845	3.3221	3.3597	3.3973
6.0	2.8536	2.8949	2.9362	2.9775	3.0189	3.0602	3.1015	3.1429
6.1	2.4997	2.5456	2.5915	2.6374	2.6833	2.7292	2.7751	2.8210
6.2	2.0549	2.1063	2.1578	2.2093	2.2608	2.3123	2.3638	2.4153
6.3	1.4963	1.5547	1.6130	1.6714	1.7297	1.7881	1.8464	1.9048
6.4	0.7949	0.8616	0.9284	0.9952	1.0620	1.1288	1.1956	1.2624
6.5	–	–	0.0670	0.1442	0.2214	0.2986	0.3758	0.4530
6.6	–	–	–	–	–	–	–	–

Table 3 (Continued)

Y	p ≐ .16	p = .17	p = .18	p = .19	p = .20	p = .21	p = .22	p = .23
2.3	–	–	–	–	–	–	–	–
2.4	–	–	–	–	–	–	–	–
2.5	–	–	–	–	–	–	–	–
2.6	9.3786	9.8251	–	–	–	–	–	–
2.7	7.9697	8.3228	8.6758	9.0288	9.3818	9.7348	–	–
2.8	6.9183	7.2002	7.4821	7.7640	8.0459	8.3278	8.6097	8.8916
2.9	6.1316	6.3589	6.5863	6.8136	7.0410	7.2683	7.4957	7.7231
3.0	5.5421	5.7273	5.9125	6.0977	6.2830	6.4682	6.6534	6.8386
3.1	5.1008	5.2532	5.4056	5.5580	5.7104	5.8628	6.0152	6.1676
3.2	4.7715	4.8982	5.0248	5.1515	5.2781	5.4048	5.5315	5.6581
3.3	4.5274	4.6337	4.7400	4.8464	4.9527	5.0590	5.1654	5.2717
3.4	4.3484	4.4386	4.5287	4.6189	4.7090	4.7992	4.8894	4.9795
3.5	4.2195	4.2967	4.3740	4.4512	4.5284	4.6056	4.6828	4.7600
3.6	4.1293	4.1960	4.2628	4.3296	4.3964	4.4632	4.5300	4.5968
3.7	4.0688	4.1271	4.1855	4.2439	4.3022	4.3606	4.4189	4.4773
3.8	4.0314	4.0829	4.1344	4.1859	4.2374	4.2889	4.3404	4.3919
3.9	4.0117	4.0576	4.1035	4.1494	4.1953	4.2412	4.2871	4.3330
4.0	4.0056	4.0469	4.0882	4.1295	4.1709	4.2122	4.2535	4.2948
4.1	4.0096	4.0472	4.0847	4.1223	4.1599	4.1975	4.2351	4.2727
4.2	4.0210	4.0555	4.0900	4.1246	4.1591	4.1936	4.2281	4.2626
4.3	4.0375	4.0695	4.1016	4.1336	4.1656	4.1976	4.2297	4.2617
4.4	4.0571	4.0871	4.1171	4.1472	4.1772	4.2072	4.2372	4.2672
4.5	4.0781	4.1065	4.1349	4.1633	4.1917	4.2201	4.2485	4.2769
4.6	4.0988	4.1260	4.1531	4.1803	4.2074	4.2346	4.2617	4.2889
4.7	4.1177	4.1439	4.1701	4.1963	4.2226	4.2488	4.2750	4.3012
4.8	4.1332	4.1588	4.1844	4.2099	4.2355	4.2611	4.2867	4.3122
4.9	4.1438	4.1690	4.1942	4.2194	4.2446	4.2698	4.2950	4.3202
5.0	4.1477	4.1728	4.1979	4.2229	4.2480	4.2731	4.2981	4.3232
5.1	4.1431	4.1683	4.1935	4.2187	4.2439	4.2691	4.2943	4.3195
5.2	4.1278	4.1534	4.1790	4.2046	4.2301	4.2557	4.2813	4.3068
5.3	4.0994	4.1256	4.1518	4.1780	4.2042	4.2305	4.2567	4.2829
5.4	4.0547	4.0819	4.1090	4.1362	4.1633	4.1905	4.2177	4.2448
5.5	3.9904	4.0188	4.0473	4.0757	4.1041	4.1325	4.1609	4.1893
5.6	3.9022	3.9322	3.9622	3.9922	4.0222	4.0523	4.0823	4.1123
5.7	3.7848	3.8168	3.8488	3.8809	3.9129	3.9449	3.9769	4.0090
5.8	3.6317	3.6662	3.7007	3.7352	3.7698	3.8043	3.8388	3.8733
5.9	3.4349	3.4724	3.5100	3.5476	3.5852	3.6228	3.6603	3.6979
6.0	3.1842	3.2255	3.2668	3.3082	3.3495	3.3908	3.4321	3.4735
6.1	2.8669	2.9128	2.9587	3.0046	3.0505	3.0964	3.1423	3.1882
6.2	2.4668	2.5183	2.5698	2.6213	2.6728	2.7243	2.7758	2.8273
6.3	1.9632	2.0215	2.0799	2.1382	2.1966	2.2549	2.3133	2.3716
6.4	1.3292	1.3960	1.4627	1.5295	1.5963	1.6631	1.7299	1.7967
6.5	0.5302	0.6074	0.6846	0.7618	0.8390	0.9163	0.9935	1.0707
6.6	–	–	–	–	–	–	0.0620	0.1522
6.7	–	–	–	–	–	–	–	–

Table 3 (Continued)

Y	p = .24	p = .25	p = .26	p = .27	p = .28	p = .29	p = .30	p = .31
2.8	9.1735	9.4554	9.7373	-	-	-	-	-
2.9	7.9504	8.1778	8.4051	8.6325	8.8599	9.0872	9.3146	9.5419
3.0	7.0238	7.2090	7.3943	7.5795	7.7647	7.9499	8.1351	8.3203
3.1	6.3200	6.4724	6.6248	6.7772	6.9296	7.0820	7.2344	7.3868
3.2	5.7848	5.9115	6.0381	6.1648	6.2914	6.4181	6.5448	6.6714
3.3	5.3780	5.4843	5.5907	5.6970	5.8033	5.9096	6.0160	6.1223
3.4	5.0697	5.1598	5.2500	5.3401	5.4303	5.5204	5.6106	5.7007
3.5	4.8372	4.9144	4.9916	5.0688	5.1461	5.2233	5.3005	5.3777
3.6	4.6636	4.7303	4.7971	4.8639	4.9307	4.9975	5.0643	5.1311
3.7	4.5356	4.5940	4.6523	4.7107	4.7690	4.8274	4.8857	4.9441
3.8	4.4434	4.4949	4.5463	4.5978	4.6493	4.7008	4.7523	4.8038
3.9	4.3789	4.4248	4.4707	4.5166	4.5625	4.6084	4.6543	4.7002
4.0	4.3362	4.3775	4.4188	4.4602	4.5015	4.5428	4.5841	4.6255
4.1	4.3102	4.3478	4.3854	4.4230	4.4606	4.4981	4.5357	4.5733
4.2	4.2972	4.3317	4.3662	4.4007	4.4352	4.4698	4.5043	4.5388
4.3	4.2937	4.3257	4.3578	4.3898	4.4218	4.4538	4.4859	4.5179
4.4	4.2972	4.3272	4.3572	4.3872	4.4172	4.4473	4.4773	4.5073
4.5	4.3053	4.3337	4.3621	4.3905	4.4189	4.4473	4.4758	4.5042
4.6	4.3160	4.3432	4.3703	4.3975	4.4246	4.4518	4.4790	4.5061
4.7	4.3274	4.3537	4.3799	4.4061	4.4323	4.4585	4.4848	4.5110
4.8	4.3378	4.3634	4.3889	4.4145	4.4401	4.4657	4.4912	4.5168
4.9	4.3453	4.3705	4.3957	4.4209	4.4461	4.4713	4.4965	4.5217
5.0	4.3483	4.3733	4.3984	4.4235	4.4485	4.4736	4.4987	4.5237
5.1	4.3447	4.3699	4.3951	4.4203	4.4454	4.4706	4.4958	4.5210
5.2	4.3324	4.3580	4.3836	4.4091	4.4347	4.4603	4.4859	4.5114
5.3	4.3091	4.3353	4.3616	4.3878	4.4140	4.4402	4.4664	4.4927
5.4	4.2720	4.2991	4.3263	4.3534	4.3806	4.4077	4.4349	4.4620
5.5	4.2177	4.2461	4.2745	4.3029	4.3313	4.3597	4.3881	4.4165
5.6	4.1423	4.1723	4.2023	4.2323	4.2623	4.2923	4.3223	4.3524
5.7	4.0410	4.0730	4.1050	4.1371	4.1691	4.2011	4.2331	4.2652
5.8	3.9078	3.9424	3.9769	4.0114	4.0459	4.0804	4.1150	4.1495
5.9	3.7355	3.7731	3.8107	3.8483	3.8858	3.9234	3.9610	3.9986
6.0	3.5148	3.5561	3.5975	3.6388	3.6801	3.7214	3.7628	3.8041
6.1	3.2341	3.2800	3.3259	3.3718	3.4178	3.4637	3.5096	3.5555
6.2	2.8788	2.9303	2.9818	3.0333	3.0848	3.1363	3.1878	3.2393
6.3	2.4300	2.4883	2.5467	2.6050	2.6634	2.7218	2.7801	2.8385
6.4	1.8635	1.9303	1.9970	2.0638	2.1306	2.1974	2.2642	2.3310
6.5	1.1479	1.2251	1.3023	1.3795	1.4567	1.5339	1.6111	1.6884
6.6	0.2423	0.3325	0.4226	0.5128	0.6029	0.6931	0.7832	0.8734
6.7	-	-	-	-	-	-	-	-

Table 3 (Continued)

Y	p = .32	p = .33	p = .34	p = .35	p = .36	p = .37	p = .38	p = .39
2.9	9.7693	9.9966	–	–	–	–	–	–
3.0	8.5055	8.6908	8.8760	9.0612	9.2464	9.4316	9.6168	9.8021
3.1	7.5392	7.6916	7.8440	7.9964	8.1488	8.3012	8.4536	8.6060
3.2	6.7981	6.9248	7.0514	7.1781	7.3047	7.4314	7.5581	7.6847
3.3	6.2286	6.3350	6.4413	6.5476	6.6539	6.7603	6.8666	6.9729
3.4	5.7909	5.8811	5.9712	6.0614	6.1515	6.2417	6.3318	6.4200
3.5	5.4549	5.5321	5.6093	5.6865	5.7637	5.8409	5.9181	5.9954
3.6	5.1979	5.2646	5.3314	5.3982	5.4650	5.5318	5.5986	5.6654
3.7	5.0025	5.0608	5.1192	5.1775	5.2359	5.2942	5.3526	5.4109
3.8	4.8553	4.9068	4.9583	5.0098	5.0613	5.1128	5.1643	5.2158
3.9	4.7461	4.7920	4.8379	4.8839	4.9298	4.9757	5.0216	5.0675
4.0	4.6668	4.7081	4.7494	4.7908	4.8321	4.8734	4.9148	4.9561
4.1	4.6109	4.6485	4.6861	4.7236	4.7612	4.7988	4.8364	4.8740
4.2	4.5733	4.6078	4.6423	4.6769	4.7114	4.7459	4.7804	4.8149
4.3	4.5499	4.5819	4.6140	4.6460	4.6780	4.7100	4.7421	4.7741
4.4	4.5373	4.5673	4.5973	4.6273	4.6573	4.6873	4.7173	4.7474
4.5	4.5326	4.5610	4.5894	4.6178	4.6462	4.6746	4.7030	4.7314
4.6	4.5333	4.5604	4.5876	4.6147	4.6419	4.6690	4.6962	4.7233
4.7	4.5372	4.5634	4.5896	4.6159	4.6421	4.6683	4.6945	4.7207
4.8	4.5424	4.5680	4.5935	4.6191	4.6447	4.6702	4.6958	4.7214
4.9	4.5469	4.5721	4.5973	4.6225	4.6476	4.6728	4.6980	4.7232
5.0	4.5488	4.5739	4.5989	4.6240	4.6491	4.6741	4.6992	4.7243
5.1	4.5462	4.5714	4.5966	4.6218	4.6470	4.6722	4.6974	4.7226
5.2	4.5370	4.5626	4.5881	4.6137	4.6393	4.6649	4.6904	4.7160
5.3	4.5189	4.5451	4.5713	4.5975	4.6238	4.6500	4.6762	4.7024
5.4	4.4892	4.5164	4.5435	4.5707	4.5978	4.6250	4.6521	4.6793
5.5	4.4449	4.4733	4.5017	4.5301	4.5585	4.5869	4.6153	4.6437
5.6	4.3824	4.4124	4.4424	4.4724	4.5024	4.5324	4.5624	4.5924
5.7	4.2972	4.3292	4.3612	4.3933	4.4253	4.4573	4.4893	4.5214
5.8	4.1840	4.2185	4.2530	4.2875	4.3221	4.3566	4.3911	4.4256
5.9	4.0362	4.0737	4.1113	4.1489	4.1865	4.2241	4.2617	4.2992
6.0	3.8454	3.8867	3.9281	3.9694	4.0107	4.0521	4.0934	4.1347
6.1	3.6014	3.6473	3.6932	3.7391	3.7850	3.8309	3.8768	3.9227
6.2	3.2908	3.3423	3.3938	3.4453	3.4968	3.5483	3.5998	3.6513
6.3	2.8968	2.9552	3.0135	3.0719	3.1302	3.1886	3.2469	3.3053
6.4	2.3978	2.4646	2.5313	2.5981	2.6649	2.7317	2.7985	2.8653
6.5	1.7656	1.8428	1.9200	1.9972	2.0744	2.1516	2.2288	2.3060
6.6	0.9635	1.0537	1.1438	1.2340	1.3242	1.4143	1.5045	1.5946
6.7	–	0.0499	0.1562	0.2626	0.3689	0.4752	0.5816	0.6879
6.8	–	–	–	–	–	–	–	–

Table 3 (Continued)

Y	p = .40	p = .41	p = .42	p = .43	p = .44	p = .45	p = .46	p = .47
3.0	9.9873	–	–	–	–	–	–	–
3.1	8.7584	8.9108	9.0633	9.2157	9.3681	9.5205	9.6729	9.8253
3.2	7.8114	7.9380	8.0647	8.1914	8.3180	8.4447	8.5714	8.6980
3.3	7.0792	7.1856	7.2919	7.3982	7.5046	7.6109	7.7172	7.8235
3.4	6.5121	6.6023	6.6924	6.7826	6.8728	6.9629	7.0531	7.1432
3.5	6.0726	6.1498	6.2270	6.3042	6.3814	6.4586	6.5358	6.6130
3.6	5.7322	5.7990	5.8657	5.9325	5.9993	6.0661	6.1329	6.1997
3.7	5.4693	5.5276	5.5860	5.6443	5.7027	5.7611	5.8194	5.8778
3.8	5.2673	5.3188	5.3703	5.4218	5.4733	5.5248	5.5763	5.6278
3.9	5.0675	5.1593	5.2052	5.2511	5.2970	5.3429	5.3888	5.4347
4.0	4.9561	5.0387	5.0801	5.1214	5.1627	5.2040	5.2454	5.2867
4.1	4.9115	4.9491	4.9867	5.0243	5.0619	5.0995	5.1370	5.1746
4.2	4.8495	4.8840	4.9185	4.9530	4.9875	5.0221	5.0566	5.0911
4.3	4.8061	4.8381	4.8702	4.9022	4.9342	4.9662	4.9983	5.0303
4.4	4.7774	4.8074	4.8374	4.8674	4.8974	4.9274	4.9574	4.9874
4.5	4.7598	4.7882	4.8166	4.8450	4.8734	4.9018	4.9302	4.9586
4.6	4.7505	4.7776	4.8048	4.8320	4.8591	4.8863	4.9134	4.9406
4.7	4.7470	4.7732	4.7994	4.8256	4.8518	4.8781	4.9043	4.9305
4.8	4.7470	4.7725	4.7981	4.8237	4.8493	4.8748	4.9004	4.9260
4.9	4.7484	4.7736	4.7988	4.8240	4.8492	4.8744	4.8996	4.9248
5.0	4.7493	4.7744	4.7995	4.8245	4.8496	4.8747	4.8997	4.9248
5.1	4.7477	4.7729	4.7981	4.8233	4.8485	4.8737	4.8989	4.9241
5.2	4.7416	4.7672	4.7927	4.8183	4.8439	4.8694	4.8950	4.9206
5.3	4.7286	4.7549	4.7811	4.8073	4.8335	4.8597	4.8860	4.9122
5.4	4.7064	4.7336	4.7607	4.7879	4.8150	4.8422	4.8694	4.8965
5.5	4.6721	4.7005	4.7289	4.7573	4.7858	4.8142	4.8426	4.8710
5.6	4.6224	4.6524	4.6825	4.7125	4.7425	4.7725	4.8025	4.8325
5.7	4.5534	4.5854	4.6174	4.6495	4.6815	4.7135	4.7455	4.7776
5.8	4.4601	4.4947	4.5292	4.5637	4.5982	4.6327	4.6673	4.7018
5.9	4.3368	4.3744	4.4120	4.4496	4.4871	4.5247	4.5623	4.5999
6.0	4.1760	4.2174	4.2587	4.3000	4.3413	4.3827	4.4240	4.4653
6.1	3.9686	4.0145	4.0604	4.1063	4.1522	4.1981	4.2440	4.2899
6.2	3.7028	3.7543	3.8057	3.8572	3.9087	3.9602	4.0117	4.0632
6.3	3.3636	3.4220	3.4803	3.5387	3.5971	3.6554	3.7138	3.7721
6.4	2.9321	2.9989	3.0657	3.1324	3.1992	3.2660	3.3328	3.3996
6.5	2.3832	2.4605	2.5377	2.6149	2.6921	2.7693	2.8465	2.9237
6.6	1.6848	1.7749	1.8651	1.9552	2.0454	2.1355	2.2257	2.3159
6.7	0.7942	0.9005	1.0069	1.1132	1.2195	1.3258	1.4322	1.5385
6.8	–	–	–	0.0354	0.1620	0.2887	0.4153	0.5420
6.9	–	–	–	–	–	–	–	–

Table 3 (Continued)

Y	p = .48	p = .49	p = .50	p = .51	p = .52	p = .53	p = .54	p = .55
3.1	9.9777	-	-	-	-	-	-	-
3.2	8.8247	8.9513	9.0780	9.2047	9.3313	9.4580	9.5847	9.7113
3.3	7.9299	8.0362	8.1425	8.2488	8.3552	8.4615	8.5678	8.6742
3.4	7.2334	7.3235	7.4137	7.5038	7.5940	7.6841	7.7743	7.8645
3.5	6.6902	6.7675	6.8447	6.9219	6.9991	7.0763	7.1535	7.2307
3.6	6.2665	6.3333	6.4000	6.4668	6.5336	6.6004	6.6672	6.7340
3.7	5.9361	5.9945	6.0528	6.1112	6.1695	6.2279	6.2862	6.3446
3.8	5.6793	5.7308	5.7823	5.8338	5.8853	5.9368	5.9883	6.0398
3.9	5.4806	5.5265	5.5724	5.6183	5.6642	5.7101	5.7560	5.8019
4.0	5.3280	5.3694	5.4107	5.4520	5.4933	5.5347	5.5760	5.6173
4.1	5.2122	5.2498	5.2874	5.3249	5.3625	5.4001	5.4377	5.4753
4.2	5.1256	5.1601	5.1947	5.2292	5.2637	5.2982	5.3327	5.3673
4.3	5.0623	5.0943	5.1264	5.1584	5.1904	5.2224	5.2545	5.2865
4.4	5.0174	5.0475	5.0775	5.1075	5.1375	5.1675	5.1975	5.2275
4.5	4.9870	5.0154	5.0438	5.0722	5.1006	5.1290	5.1574	5.1858
4.6	4.9677	4.9949	5.0220	5.0492	5.0763	5.1035	5.1306	5.1578
4.7	4.9567	4.9829	5.0092	5.0354	5.0616	5.0878	5.1140	5.1403
4.8	4.9515	4.9771	5.0027	5.0283	5.0538	5.0794	5.1050	5.1306
4.9	4.9500	4.9751	5.0003	5.0255	5.0507	5.0759	5.1011	5.1263
5.0	4.9499	4.9749	5.0000	5.0251	5.0501	5.0752	5.1003	5.1253
5.1	4.9493	4.9745	4.9997	5.0249	5.0500	5.0752	5.1004	5.1256
5.2	4.9462	4.9717	4.9973	5.0229	5.0485	5.0740	5.0996	5.1252
5.3	4.9384	4.9646	4.9908	5.0171	5.0433	5.0695	5.0957	5.1219
5.4	4.9237	4.9508	4.9780	5.0051	5.0323	5.0594	5.0866	5.1137
5.5	4.8994	4.9278	4.9562	4.9846	5.0130	5.0414	5.0698	5.0982
5.6	4.8625	4.8925	4.9225	4.9525	4.9826	5.0126	5.0426	5.0726
5.7	4.8096	4.8416	4.8736	4.9057	4.9377	4.9697	5.0017	5.0338
5.8	4.7363	4.7708	4.8053	4.8399	4.8744	4.9089	4.9434	4.9779
5.9	4.6375	4.6751	4.7126	4.7502	4.7878	4.8254	4.8630	4.9005
6.0	4.5067	4.5480	4.5893	4.6306	4.6720	4.7133	4.7546	4.7960
6.1	4.3358	4.3817	4.4276	4.4735	4.5194	4.5653	4.6112	4.6571
6.2	4.1147	4.1662	4.2177	4.2692	4.3207	4.3722	4.4237	4.4752
6.3	3.8305	3.8888	3.9472	4.0055	4.0639	4.1222	4.1806	4.2389
6.4	3.4664	3.5332	3.6000	3.6667	3.7335	3.8003	3.8671	3.9339
6.5	3.0009	3.0781	3.1553	3.2325	3.3098	3.3870	3.4642	3.5414
6.6	2.4060	2.4962	2.5863	2.6765	2.7666	2.8568	2.9469	3.0371
6.7	1.6448	1.7512	1.8575	1.9638	2.0701	2.1765	2.2828	2.3891
6.8	0.6687	0.7953	0.9220	1.0487	1.1753	1.3020	1.4286	1.5553
6.9	-	-	-	-	0.0223	0.1747	0.3271	0.4795
7.0	-	-	-	-	-	-	-	-

Table 3 (Continued)

Y	p = .56	p = .57	p = .58	p = .59	p = .60	p = .61	p = .62	p = .63
3.2	9.8380	9.9646	–	–	–	–	–	–
3.3	8.7805	8.8868	8.9931	9.0995	9.2058	9.3121	9.4184	9.5248
3.4	7.9546	8.0448	8.1349	8.2251	8.3152	8.4054	8.4955	8.5857
3.5	7.3079	7.3851	7.4623	7.5395	7.6168	7.6940	7.7712	7.8484
3.6	6.8008	6.8676	6.9343	7.0011	7.0679	7.1347	7.2015	7.2683
3.7	6.4029	6.4613	6.5197	6.5780	6.6364	6.6947	6.7531	6.8114
3.8	6.0913	6.1428	6.1943	6.2457	6.2972	6.3487	6.4002	6.4517
3.9	5.8478	5.8937	5.9396	5.9855	6.0314	6.0773	6.1232	6.1691
4.0	5.6587	5.7000	5.7413	5.7826	5.8240	5.8653	5.9066	5.9479
4.1	5.5129	5.5504	5.5880	5.6256	5.6632	5.7008	5.7383	5.7759
4.2	5.4018	5.4363	5.4708	5.5053	5.5399	5.5744	5.6089	5.6434
4.3	5.3185	5.3505	5.3826	5.4146	5.4466	5.4786	5.5107	5.5427
4.4	5.2575	5.2875	5.3175	5.3476	5.3776	5.4076	5.4376	5.4676
4.5	5.2142	5.2427	5.2711	5.2995	5.3279	5.3563	5.3847	5.4131
4.6	5.1850	5.2121	5.2393	5.2664	5.2936	5.3207	5.3479	5.3750
4.7	5.1665	5.1927	5.2189	5.2451	5.2714	5.2976	5.3238	5.3500
4.8	5.1561	5.1817	5.2073	5.2328	5.2584	5.2840	5.3096	5.3351
4.9	5.1515	5.1767	5.2019	5.2271	5.2523	5.2774	5.3026	5.3278
5.0	5.1504	5.1755	5.2005	5.2256	5.2507	5.2757	5.3008	5.3259
5.1	5.1508	5.1760	5.2012	5.2264	5.2516	5.2768	5.3020	5.3272
5.2	5.1507	5.1763	5.2019	5.2275	5.2530	5.2786	5.3042	5.3298
5.3	5.1482	5.1744	5.2006	5.2268	5.2530	5.2793	5.3055	5.3317
5.4	5.1409	5.1680	5.1952	5.2224	5.2495	5.2767	5.3038	5.3310
5.5	5.1266	5.1550	5.1834	5.2118	5.2402	5.2686	5.2970	5.3254
5.6	5.1026	5.1326	5.1626	5.1926	5.2226	5.2526	5.2827	5.3127
5.7	5.0658	5.0978	5.1298	5.1619	5.1939	5.2259	5.2579	5.2900
5.8	5.0125	5.0470	5.0815	5.1160	5.1505	5.1851	5.2196	5.2541
5.9	5.9381	4.9757	5.0133	5.0509	5.0885	5.1260	5.1636	5.2012
6.0	4.8373	4.8786	4.9199	4.9613	5.0026	5.0439	5.0852	5.1266
6.1	4.7030	4.7489	4.7948	4.8407	4.8866	4.9325	4.9784	5.0243
6.2	4.5267	4.5782	4.6297	4.6812	4.7327	4.7842	4.8357	4.8872
6.3	4.2973	4.3557	4.4140	4.4724	4.5307	4.5891	4.6474	4.7058
6.4	4.0007	4.0675	4.1343	4.2010	4.2678	4.3346	4.4014	4.4682
6.5	3.6186	3.6958	3.7730	3.8502	3.9274	4.0046	4.0819	4.1591
6.6	3.1272	3.2174	3.3076	3.3977	3.4879	3.5780	3.6682	3.7583
6.7	2.4954	2.6018	2.7081	2.8144	2.9208	3.0271	3.1334	3.2397
6.8	1.6820	1.8086	1.9353	2.0620	2.1886	2.3153	2.4419	2.5686
6.9	0.6319	0.7863	0.9367	1.0892	1.2416	1.3940	1.5464	1.6988
7.0	–	–	–	–	0.0127	.1979	0.3832	0.5684

Table 3 (Continued)

Y	p = .64	p = .65	p = .66	p = .67	p = .68	p = .69	p = .70	p = .71
3.3	9.6311	9.7374	9.8438	9.9501	–	–	–	–
3.4	8.6758	8.7660	8.8562	8.9463	9.0365	9.1266	9.2168	9.3069
3.5	7.9256	8.0028	8.0800	8.1572	8.2344	8.3116	8.3889	8.4661
3.6	7.3351	7.4019	7.4687	7.5354	7.6022	7.6690	7.7358	7.8026
3.7	6.8698	6.9281	6.9865	7.0448	7.1032	7.1615	7.2199	7.2782
3.8	6.5032	6.5547	6.6062	6.6577	6.7092	6.7607	6.8122	6.8637
3.9	6.2150	6.2609	6.3068	6.3527	6.3986	6.4445	6.4904	6.5363
4.0	5.9893	6.0306	6.0719	6.1133	6.1546	6.1959	6.2372	6.2786
4.1	5.8135	5.8511	5.8887	5.9263	5.9638	6.0014	6.0390	6.0766
4.2	5.6779	5.7125	5.7470	5.7815	5.8160	5.8505	5.8850	5.9196
4.3	5.5747	5.6067	5.6388	5.6708	5.7028	5.7348	5.7669	5.7989
4.4	5.4976	5.5276	5.5576	5.5876	5.6176	5.6476	5.6777	5.7077
4.5	5.4415	5.4699	5.4983	5.5267	5.5551	5.5835	5.6119	5.6403
4.6	5.4022	5.4293	5.4565	5.4836	5.5108	5.5380	5.5651	5.5923
4.7	5.3762	5.4025	5.4287	5.4549	5.4811	5.5073	5.5336	5.5598
4.8	5.3607	5.3863	5.4119	5.4374	5.4630	5.4886	5.5141	5.5397
4.9	5.3530	5.3782	5.4034	5.4286	5.4538	5.4790	5.5042	5.5294
5.0	5.3509	5.3760	5.4011	5.4261	5.4512	5.4763	5.5013	5.5264
5.1	5.3524	5.3775	5.4027	5.4279	5.4531	5.4783	5.5035	5.5287
5.2	5.3553	5.3809	5.4065	5.4320	5.4576	5.4832	5.5088	5.5343
5.3	5.3579	5.3841	5.4104	5.4366	5.4628	5.4890	5.5152	5.5415
5.4	5.3581	5.3853	5.4124	5.4396	5.4667	5.4939	5.5210	5.5482
5.5	5.3538	5.3822	5.4106	5.4390	5.4674	5.4958	5.5242	5.5527
5.6	5.3427	5.3727	5.4027	5.4327	5.4627	5.4927	5.5227	5.5527
5.7	5.3220	5.3540	5.3860	5.4181	5.4501	5.4821	5.5141	5.5462
5.8	5.2886	5.3231	5.3577	5.3922	5.4267	5.4612	5.4957	5.5302
5.9	5.2388	5.2764	5.3139	5.3515	5.3891	5.4267	5.4643	5.5019
6.0	5.1679	5.2092	5.2506	5.2919	5.3332	5.3745	5.4159	5.4572
6.1	5.0702	5.1161	5.1621	5.2080	5.2539	5.2998	5.3457	5.3916
6.2	4.9387	4.9902	5.0417	5.0932	5.1447	5.1962	5.2477	5.2992
6.3	4.7641	4.8225	4.8808	4.9392	4.9975	5.0559	5.1143	5.1726
6.4	4.5350	4.6018	4.6686	4.7354	4.8021	4.8689	4.9357	5.0025
6.5	4.2363	4.3135	4.3907	4.4679	4.5451	4.6223	4.6995	4.7767
6.6	3.8485	3.9386	4.0288	4.1189	4.2091	4.2993	4.3894	4.4796
6.7	3.3461	3.4524	3.5587	3.6650	3.7714	3.8777	3.9840	4.0904
6.8	2.6953	2.8219	2.9486	3.0752	3.2019	3.3286	3.4552	3.5819
6.9	1.8512	2.0036	2.1560	2.3084	2.4608	2.6132	2.7656	2.9180
7.0	0.7536	0.9388	1.1240	1.3092	1.4945	1.6797	1.8649	2.0501
7.1	–	–	–	0.0034	0.2307	0.4581	0.6854	0.9128

Table 3 (Continued)

Y	p = .72	p = .73	p = .74	p = .75	p = .76	p = .77	p = .78	p = .79
3.4	9.3971	9.4872	9.5774	9.6675	9.7577	9.8478	9.9380	–
3.5	8.5433	8.6205	8.6977	8.7749	8.8521	8.9293	9.0065	9.0837
3.6	7.8694	7.9362	8.0030	8.0697	8.1365	8.2033	8.2701	8.3369
3.7	7.3366	7.3950	7.4533	7.5117	7.5700	7.6284	7.6867	7.7451
3.8	6.9152	6.9667	7.0182	7.0697	7.1212	7.1727	7.2242	7.2757
3.9	6.5822	6.6282	6.6741	6.7200	6.7659	6.8118	6.8577	6.9036
4.0	6.3199	6.3612	6.4025	6.4439	6.4852	6.5265	6.5679	6.6092
4.1	6.1142	6.1517	6.1893	6.2269	6.2645	6.3021	6.3397	6.3772
4.2	5.9541	5.9886	6.0231	6.0576	6.0922	6.1267	6.1612	6.1957
4.3	5.8309	5.8629	5.8950	5.9270	5.9590	5.9910	6.0231	6.0551
4.4	5.7377	5.7677	5.7977	5.8277	5.8577	5.8877	5.9177	5.9477
4.5	5.6687	5.6971	5.7255	5.7539	5.7823	5.8107	5.8391	5.8675
4.6	5.6194	5.6466	5.6737	5.7009	5.7280	5.7552	5.7823	5.8095
4.7	5.5860	5.6122	5.6384	5.6647	5.6909	5.7171	5.7433	5.7695
4.8	5.5653	5.5909	5.6164	5.6420	5.6676	5.6932	5.7187	5.7443
4.9	5.5546	5.5797	5.6049	5.6301	5.6553	5.6805	5.7057	5.7309
5.0	5.5515	5.5765	5.6016	5.6267	5.6517	5.6768	5.7019	5.7269
5.1	5.5539	5.5791	5.6043	5.6295	5.6547	5.6798	5.7050	5.7302
5.2	5.5599	5.5855	5.6111	5.6366	5.6622	5.6878	5.7133	5.7389
5.3	5.5677	5.5939	5.6201	5.6463	5.6726	5.6988	5.7250	5.7512
5.4	5.5754	5.6025	5.6297	5.6568	5.6840	5.7111	5.7383	5.7654
5.5	5.5811	5.6095	5.6379	5.6663	5.6947	5.7231	5.7515	5.7799
5.6	5.5828	5.6128	5.6428	5.6728	5.7028	5.7328	5.7628	5.7928
5.7	5.5782	5.6102	5.6422	5.6743	5.7063	5.7383	5.7703	5.8024
5.8	5.5648	5.5993	5.6338	5.6683	5.7028	5.7374	5.7719	5.8064
5.9	5.5394	5.5770	5.6146	5.6522	5.6898	5.7273	5.7649	5.8025
6.0	5.4985	5.5398	5.5812	5.6225	5.6638	5.7052	5.7465	5.7878
6.1	5.4375	5.4834	5.5293	5.5752	5.6211	5.6670	5.7129	5.7588
6.2	5.3507	5.4022	5.4537	5.5051	5.5566	5.6081	5.6596	5.7111
6.3	5.2310	5.2893	5.3477	5.4060	5.4644	5.5227	5.5811	5.6394
6.4	5.0693	5.1361	5.2029	5.2697	5.3364	5.4032	5.4700	5.5368
6.5	4.8539	4.9312	5.0084	5.0856	5.1628	5.2400	5.3172	5.3944
6.6	4.5697	4.6599	4.7500	4.8402	4.9303	5.0205	5.1106	5.2008
6.7	4.1967	4.3030	4.4093	4.5157	4.6220	4.7283	4.8346	4.9410
6.8	3.7086	3.8352	3.9619	4.0885	4.2152	4.3419	4.4685	4.5952
6.9	3.0704	3.2228	3.3752	3.5276	3.6800	3.8324	3.9848	4.1372
7.0	2.2353	2.4205	2.6057	2.7910	2.9762	3.1614	3.3466	3.5318
7.1	1.1401	1.3675	1.5949	1.8222	2.0496	2.2769	2.5043	2.7317
7.2	–	–	0.2627	0.5446	0.8265	1.1084	1.3903	1.6722
7.3	–	–	–	–	–	–	–	0.2652

Table 3 (Continued)

Y	p = .80	p = .81	p = .82	p = .83	p = .84	p = .85	p = .86	p = .87
3.5	9.1610	9.2382	9.3154	9.3926	9.4698	9.5470	9.6242	9.7014
3.6	8.4037	8.4705	8.5373	8.6040	8.6708	8.7376	8.8044	8.8712
3.7	7.8034	7.8618	7.9201	7.9785	8.0368	8.0952	8.1536	8.2119
3.8	7.3272	7.3787	7.4302	7.4817	7.5332	7.5847	7.6362	7.6877
3.9	6.9495	6.9954	7.0413	7.0872	7.1331	7.1790	7.2249	7.2708
4.0	6.6505	6.6918	6.7332	6.7745	6.8158	6.8571	6.8985	6.9398
4.1	6.4148	6.4524	6.4900	6.5276	6.5651	6.6027	6.6403	6.6779
4.2	6.2302	6.2648	6.2993	6.3338	6.3683	6.4028	6.4374	6.4719
4.3	6.0871	6.1191	6.1512	6.1832	6.2152	6.2473	6.2793	6.3113
4.4	5.9778	6.0078	6.0378	6.0678	6.0978	6.1278	6.1578	6.1878
4.5	5.8959	5.9243	5.9527	5.9812	6.0096	6.0380	6.0664	6.0948
4.6	5.8367	5.8638	5.8910	5.9181	5.9453	5.9724	5.9996	6.0267
4.7	5.7958	5.8220	5.8482	5.8744	5.9006	5.9269	5.9531	5.9793
4.8	5.7699	5.7954	5.8210	5.8466	5.8722	5.8977	5.9233	5.9489
4.9	5.7561	5.7813	5.8065	5.8317	5.8569	5.8821	5.9072	5.9324
5.0	5.7520	5.7771	5.8021	5.8272	5.8523	5.8773	5.9024	5.9275
5.1	5.7554	5.7806	5.8058	5.8310	5.8562	5.8814	5.9066	5.9318
5.2	5.7645	5.7901	5.8156	5.8412	5.8668	5.8924	5.9179	5.9435
5.3	5.7774	5.8037	5.8299	5.8561	5.8823	5.9085	5.9348	5.9610
5.4	5.7926	5.8197	5.8469	5.8740	5.9012	5.9284	5.9555	5.9827
5.5	5.8083	5.8367	5.8651	5.8935	5.9219	5.9503	5.9787	6.0071
5.6	5.8228	5.8528	5.8829	5.9129	5.9429	5.9729	6.0029	6.0329
5.7	5.8344	5.8664	5.8984	5.9305	5.9625	5.9945	6.0265	6.0586
5.8	5.8409	5.8754	5.9100	5.9445	5.9790	6.0135	6.0480	6.0826
5.9	5.8401	5.8777	5.9153	5.9528	5.9904	6.0280	6.0656	6.1032
6.0	5.8291	5.8705	5.9118	5.9531	5.9944	6.0358	6.0771	6.1184
6.1	5.8047	5.8506	5.8965	5.9424	5.9883	6.0342	6.0801	6.1260
6.2	5.7626	5.8141	5.8656	5.9171	5.9686	6.0201	6.0716	6.1231
6.3	5.6978	5.7561	5.8145	5.8729	5.9312	5.9896	6.0479	6.1063
6.4	5.6036	5.6704	5.7372	5.8040	5.8707	5.9375	6.0043	6.0711
6.5	5.4716	5.5488	5.6260	5.7033	5.7805	5.8577	5.9349	6.0121
6.6	5.2910	5.3811	5.4713	5.5614	5.6516	5.7417	5.8319	5.9220
6.7	5.0473	5.1536	5.2600	5.3663	5.4726	5.5789	5.6853	5.7916
6.8	4.7219	4.8485	4.9752	5.1018	5.2285	5.3552	5.4818	5.6085
6.9	4.2896	4.4420	4.5944	4.7468	4.8992	5.0516	5.2040	5.3564
7.0	3.7170	3.9023	4.0875	4.2727	4.4579	4.6431	4.8283	5.0136
7.1	2.9590	3.1864	3.4137	3.6411	3.8684	4.0958	4.3232	4.5505
7.2	1.9541	2.2360	2.5179	2.7998	3.0817	3.3635	3.6454	3.9273
7.3	0.6182	0.9712	1.3242	1.6772	2.0303	2.3833	2.7363	3.0893
7.4	-	-	-	0.1749	0.6214	1.0680	1.5145	1.9611
7.5	-	-	-	-	-	-	-	0.4377

Table 3 (Continued)

Y	p = .88	p = .89	p = .90	p = .91	p = 92	p = .93	p = .94	p = .95
3.5	9.7786	9.8558	9.9330	–	–	–	–	–
3.6	8.9380	9.0048	9.0716	9.1384	9.2051	9.2719	9.3387	9.4055
3.7	8.2703	8.3286	8.3870	8.4453	8.5037	8.5620	8.6204	8.6787
3.8	7.7392	7.7907	7.8422	7.8937	7.9451	7.9966	8.0481	8.0996
3.9	7.3167	7.3626	7.4085	7.4544	7.5003	7.5462	7.5921	7.6380
4.0	6.9811	7.0225	7.0638	7.1051	7.1464	7.1878	7.2291	7.2704
4.1	6.7155	6.7531	6.7906	6.8282	6.8658	6.9034	6.9410	6.9786
4.2	6.5064	6.5409	6.5754	6.6100	6.6445	6.6790	6.7135	6.7480
4.3	6.3433	6.3754	6.4074	6.4394	6.4714	6.5035	6.5355	**6.5675**
4.4	6.2178	6.2478	6.2779	6.3079	6.3379	6.3679	6.3979	6.4279
4.5	6.1232	6.1516	6.1800	6.2084	6.2368	6.2652	6.2936	6.3220
4.6	6.0539	6.0810	6.1082	6.1353	6.1625	6.1897	6.2168	6.2440
4.7	6.0055	6.0317	6.0580	6.0842	6.1104	6.1366	6.1628	6.1891
4.8	5.9744	6.0000	6.0256	6.0512	6.0767	6.1023	6.1279	6.1535
4.9	5.9576	5.9828	6.0080	6.0332	6.0584	6.0836	6.1088	6.1340
5.0	5.9525	5.9776	6.0027	6.0277	6.0528	6.0779	6.1029	6.1280
5.1	5.9570	5.9822	6.0073	6.0325	6.0577	6.0829	6.1081	6.1333
5.2	5.9691	5.9946	6.0202	6.0458	6.0714	6.0969	6.1225	6.1481
5.3	5.9872	6.0134	6.0396	6.0659	6.0921	6.1183	6.1445	6.1707
5.4	6.0098	6.0370	6.0641	6.0913	6.1184	6.1456	6.1727	6.1999
5.5	6.0355	6.0639	6.0923	6.1207	6.1491	6.1775	6.2059	6.2343
5.6	6.0629	6.0929	6.1229	6.1529	6.1829	6.2130	6.2430	6.2730
5.7	6.0906	6.1226	6.1546	6.1867	6.2187	6.2507	6.2827	6.3148
5.8	6.1171	6.1516	6.1861	6.2206	6.2552	6.2897	6.3242	6.3587
5.9	6.1408	6.1783	6.2159	6.2535	6.2911	6.3287	6.3662	6.4038
6.0	6.1598	6.2011	6.2424	6.2837	6.3251	6.3664	6.4077	6.4490
6.1	6.1719	6.2178	6.2637	6.3096	6.3555	6.4014	6.4473	6.4932
6.2	6.1746	6.2261	6.2776	6.3291	6.3806	6.4321	6.4836	6.5351
6.3	6.1646	6.2230	6.2813	6.3397	6.3980	6.4564	6.5147	6.5731
6.4	6.1379	6.2047	6.2715	6.3383	6.4051	6.4718	6.5386	6.6054
6.5	6.0893	6.1665	6.2437	6.3209	6.3981	6.4753	6.5526	6.6298
6.6	6.0122	6.1023	6.1925	6.2827	6.3728	6.4630	6.5531	6.6433
6.7	5.8979	6.0043	6.1106	6.2169	6.3232	6.4296	6.5359	6.6422
6.8	5.7352	5.8618	5.9885	6.1151	6.2418	6.3685	6.4951	6.6218
6.9	5.5088	5.6612	5.8136	5.9660	6.1184	6.2708	6.4232	6.5756
7.0	5.1988	5.3840	5.5692	5.7544	5.9396	6.1249	6.3101	6.4953
7.1	4.7779	5.0052	5.2326	5.4599	5.6873	5.9147	6.1420	6.3694
7.2	4.2092	4.4911	4.7730	5.0549	5.3368	5.6187	5.9006	6.1825
7.3	3.4423	3.7954	4.1484	4.5014	4.8544	5.2074	5.5605	5.9135
7.4	2.4076	2.8541	3.3007	3.7472	4.1937	4.6403	5.0868	5.5334
7.5	1.0082	1.5787	2.1492	2.7197	3.2902	3.8607	4.4312	5.0017
7.6	–	–	0.5810	1.3172	2.0534	2.7897	3.5259	4.2621
7.7	–	–	–	–	0.3558	1.3155	2.2751	3.2347
7.8	–	–	–	–	–	–	0.5427	1.8061

Table 3 (Continued)

Y	p = .96	p = .97	p = .98	p = .99	p = 1.00	Y	p = 1.00
3.6	9.4723	9.5391	9.6059	9.6727	9.7394	8.5	8.7666
3.7	8.7371	8.7954	8.8538	8.9122	8.9705	8.6	8.8601
3.8	8.1511	8.2026	8.2541	8.3056	8.3571	8.7	9.9538
3.9	7.6839	7.7298	7.7757	7.8216	7.8675	8.8	9.0479
4.0	7.3117	7.3531	7.3944	7.4357	7.4771	8.9	9.1422
						9.0	9.2368
4.1	7.0161	7.0537	7.0913	7.1289	7.1665		
4.2	6.7826	6.8171	6.8516	6.8861	6.9206	9.1	9.3316
4.3	6.5995	6.6316	6.6636	6.6956	6.7276	9.2	9.4266
4.4	6.4579	6.4879	6.5179	6.5479	6.5780	9.3	9.5218
4.5	6.3504	6.3788	6.4072	6.4356	6.4640	9.4	9.6172
4.6	6.2711	6.2983	6.3254	6.3254	6.3797	9.5	9.7128
4.7	6.2153	6.2415	6.2677	6.2677	6.3202	9.6	9.8085
4.8	6.1790	6.2046	6.2302	6.2302	6.2813	9.7	9.9045
4.9	6.1592	6.1844	6.2095	6.2095	6.2599		
5.0	6.1530	6.1781	6.2032	6.2032	6.2533		
5.1	6.1585	6.1837	6.2089	6.2089	6.2593		
5.2	6.1737	6.1992	6.2248	6.2248	6.2759		
5.3	6.1970	6.2232	6.2494	6.2494	6.3018		
5.4	6.2271	6.2542	6.2814	6.2814	6.3357		
5.5	6.2627	6.2912	6.3196	6.3196	6.3764		
5.6	6.3030	6.3330	6.3630	6.3630	6.4230		
5.7	6.3468	6.3788	6.4108	6.4108	6.4749		
5.8	6.3932	6.4278	6.4623	6.4623	6.5313		
5.9	6.4414	6.4790	6.5166	6.5166	6.5917		
6.0	6.4904	6.5317	6.5730	6.5730	6.6557		
6.1	6.5391	6.5850	6.6309	6.6768	6.7227		
6.2	6.5866	6.6381	6.6896	6.7411	6.7927		
6.3	6.6315	6.6898	6.7482	6.8065	6.8649		
6.4	6.6722	6.7390	6.8058	6.8726	6.9394		
6.5	6.7070	6.7842	6.8614	6.9386	7.0158		
6.6	6.7334	6.8236	6.9137	7.0039	7.0940		
6.7	6.7485	6.8549	6.9612	7.0675	7.1739		
6.8	6.7485	6.8751	7.0018	7.1284	7.2551		
6.9	6.7280	6.8804	7.0328	7.1852	7.3376		
7.0	6.6805	6.8657	7.0509	7.2362	7.4214		
7.1	6.3694	6.8241	7.0514	7.2788	7.5062		
7.2	6.1825	6.7462	7.0281	7.3100	7.5919		
7.3	6.2665	6.6195	6.9725	7.3256	7.6786		
7.4	5.9799	6.4264	6.8730	7.3195	7.7661		
7.5	5.5722	6.1427	6.7133	7.2838	7.8543		
7.6	4.9983	5.7345	6.4707	7.2070	7.9432		
7.7	4.1943	5.1539	6.1135	7.0731	8.0327		
7.8	3.9694	4.3328	5.5961	6.8595	8.1228		
7.9	1.4936	3.1736	4.8535	6.5335	8.2135		
8.0	–	1.5354	3.7918	6.0482	8.3046		
8.1	–	–	2.2740	5.3351	8.3962		
8.2	–	–	0.0993	4.2938	8.4882		
8.3	–	–	–	2.7754	8.5807		
8.4	–	–	–	0.5580	8.6735		

Bibliography and Cited References

Index to Bibliography

The number after an entry corresponds to the listing in the Bibliography which contains 766 references related to Bioassay. It is interesting to note that 60% of the references are dated within the last 15 years and this reflects the advances in Bioassay research. Only the primary purpose of the reference is the reason for its classification, even though a reference may have multiple purposes. Admittedly, this is arbitrary and is my subjective opinion. Occasionally, however, a reference may be in two places if the primary and secondary purposes are not distinguishable. This Index to the Bibliography is a response to reviewers to whom I am grateful.

The 7 main divisions of the index are:

PART 1: Univariate Methods

Combination of Assays: 25, 28–30, 44, 70, 72, 75, 464, 606, 719, 742
Dilution Assays: 24, 31, 312, 415, 490, 516, 517, 622, 630, 760
Direct Assays: 34, 110, 622, 623, 630
Joint Action: 3, 4, 38, 40, 99, 235, 367, 612, 627, 657, 658, 721
Mixtures: 36, 42, 43, 46, 136, 280, 368, 369, 371, 372, 373, 374, 393, 433, 454, 492, 551, 552, 553
Parabolic Assays: 104, 203, 261, 723, 740, 751
Parallel-line assays: 164, 198, 218, 305, 495, 540, 572, 604, 637, 709, 716, 746, 756
Quantal Assays:
    Methods: 7, 27, 37, 41, 125, 129, 130, 131, 186, 197, 344, 370, 385, 407, 408, 410, 507, 745, 747
    ED50 Estimation: see Part 2.
Radioimmunoassay (RIA):
    Data Processing: 193, 540, 591, 592, 597
    Models: 107, 160, 296, 315, 322, 329, 349, 593, 595, 596, 604, 691
    Estimation: 11, 304, 421, 460, 489, 563, 588, 691, 692, 698, 715
    Theoretical Aspects: 167, 298, 360, 500, 546, 547, 589, 590, 594, 639, 715, 752, 753
Slope-ratio assays: 71, 184, 580, 716
Time-dependency in Bioassay:
    Life Table Approach: 430
    Median Survival Time: 124, 172, 173, 263, 466, 481, 575, 636
    Quantal Assays: 52, 97, 151, 156, 389, 443, 734
    Time to Response Models: 418, 453, 607

PART 2: ED50 Estimation Methods

Bayesian: 240, 442, 669
Binomial or Binary Aspects: 4, 22, 23, 452
Distribution-free: 633, 707
Dixon-Mood: 243
Extra-binomial Variation: 183, 202, 531, 620, 743
Logits: 10, 15, 48, 76, 77, 81, 82, 83, 86, 87, 133, 254, 291, 324, 499, 642, 643, 693, 703, 758
Low dose: 326, 327, 354, 554, 605, 618, 628, 635
Moving Average: 67, 350, 686, 687
Non-parametric: 180, 680, 573
Normits, 85
Probits: 8, 10, 14, 21, 61, 65, 94, 95, 96, 170, 171, 356, 365, 521, 556, 634, 679, 694, 757
Quantits: 194, 195, 518, 557

Ranks: 74
Reed-Muench: 88, 137, 550, 574
Ridits: 352
Robbins-Munro: 188, 579
Sequential: 189, 190, 423, 525 (See also Part 5 below.)
Spearman-Karber: 126, 127, 166, 176, 181, 347, 348, 415, 420, 509, 561, 650
Tobits: 392, 524
Up-and-down: 138, 177, 241, 244, 390, 424, 472, 473, 701

PART 3: Multivariate Methods

Combination of Assays: 719
Growth Curves: 53, 116, 155, 253, 341, 542, 555, 569, 577, 669, 763, 764
Parabolic: 159, 723, 724
Parallel-line: 154, 156, 157, 158, 238, 455, 717, 765
Quantal: 45, 151, 152, 153, 156, 394, 396, 429, 432, 718
Slope-ratio: 403
Theoretical Papers: 68, 69, 73, 134, 321, 351, 425, 426, 427, 441, 442, 461, 533, 567, 568, 600, 651
Theoretical Texts: 92, 311, 661, 622

PART 4: Computer Related

Manuals: 57, 66, 211, 231, 237, 242, 331, 340, 549, 598, 641, 675
Software:
    Probit Analysis: 16, 220, 310, 399, 401, 402, 534, 583, 601, 602, 619
    Parabolic: 723
    Logits: 267, 440, 583
    Parallel-line: 297, 397, 482, 495, 540, 604, 624
    Growth Curves: 13
    RIA: 11, 540, 597
    Slope-ratio: 398, 562

PART 5: Theoretical or Mathematical Related Papers to Bioassay

Approximations: 19, 77, 84, 191, 480, 514, 532, 579, 616, 672, 711
Bayesian: 12, 442, 451, 566, 669
    Quantal: 17, 18, 63, 88, 240, 314, 404
    Quantitative: 218
Design of Experiments: 2, 105, 106, 108, 117, 165, 199, 200, 219, 286, 287, 290, 293, 299, 301, 359, 405, 406, 447, 448, 491, 578, 637, 700, 702
Distribution Theory: 50, 162, 175, 206, 250, 312, 434, 663, 684, 736, 737
Graphical Methods: 232, 233, 466, 467, 469, 559, 630
Inverse Regression: 345, 444, 445, 459, 313, 631, 632, 744
Jackknife: 32, 93, 121, 269, 313, 337, 338, 361, 508, 510, 511, 605, 625, 626, 689
Models: 163, 196, 207, 384, 446, 452
Repeated Measures Design: 6, 258, 303, 640
Risk Assessment: 201
Sampling: 59
Sequential Theory: 227–229, 256, 457, 586, 660, 731–733, 750
Survival Analysis: 681
Testing: 51, 414, 629, 676
Transformations: 62, 78, 90, 179, 185, 343, 431, 526, 688
Theoretical Aspects of Estimation
    Confidence Limits: 210, 213, 273–276, 353, 438, 439, 470, 484, 571
    Control or Zero dose: 114, 288, 307, 738, 739
    Covariance: 294
    Grouped Data: 317, 318
    Error Aspects: 80, 289, 538, 648, 649, 665–667, 696, 708, 730
    Maximum Likelihood: 200, 325, 407, 417, 458, 488, 530, 617
    Ratios: 119, 187, 209, 272, 279, 328, 376, 377, 438, 484, 501, 503, 520, 611, 671
    Regression Methods: 20, 79
    Restricted Estimation: 139
    Robust Estimation: 346, 512, 570

Serial Correlation: 216, 683
Small Sample Estimation: 182, 486, 705
Tables: 5, 85, 306, 727, 749
Two-sample Methods: 33

PART 6: Examples Illustrating Methods

Agriculture: 60, 148, 364, 587; Chickens: 613; Cows: 335, 535; Tobacco: 614, 615; Tomatoes: 478, 493, 494
Anthropology: 278, 576
Aquatic Toxicology:
    Acute: 55, 136, 382, 436, 603, 653, 654, 656
    Basic Concepts: 135, 260, 357, 358, 381, 383, 479, 485, 628
    Behaviour: 316
    Growth: 437
    Subacute: 246, 247, 655
    Threshold Value: 142, 168, 169, 178
    Agent: Ammonia: 123, 136, 690; Cadmium: 496; Chemicals: 55, 264, 433, 697; Chlorine: 582; Copper: 255, 465, 498, 652; Cyanide: 123; Lead: 380; Metal Ions: 115, 248; Oil: 657; Permethrin: 449; Selenium: 529; Temperature: 120, 382; Zinc: 123, 136, 255, 379, 474, 658
    Subject: Algae: 477, 748; Fathead Minnow: 123; Fauna: 766; Frogs: 608; Goldfish: 496; Lake Trout: 464; Lobster: 497, 498; Rainbow Trout: 123, 136, 380, 449, 465, 474, 603, 712, 726; Salmon: 120, 225, 379, 380, 652, 658; Zebrafish: 529
Cancer: 212, 213, 214, 461, 487, 543, 544, 545, 645, 646
Coal Mining: 39
Cosmetics: 9
Forestry: 584, 660, 710
Genetics: 722, 741
Industrial: 362, 729
Insecticides: 1, 101, 122, 140, 144, 146, 150, 223, 282, 284, 309, 387, 388, 506, 523, 536, 548, 558, 584, 585, 609, 610, 668, 670, 677, 678, 682, 706, 713, 725, 735
Insulin: 111, 272, 277, 755
Medicine: 35, 64, 91, 102, 141, 161, 245, 308, 334, 363, 409, 416, 492, 522
Menarche: 112, 224, 251, 323, 336, 412, 462, 581
Meteorology: 54
Nutrition: 695, 699
Pesticides for:
    Beetles: 309, 366, 367, 599; Bees: 49, 281; Birds: 252; Flies: 100, 113; Mice: 644; Pheasants: 330; Rabbits: 249; Rats: 673; Weeds: 386
Pharmacology: 58, 98, 109, 132, 208, 230, 283, 285, 320, 332, 339, 355, 391, 395, 411, 413, 428, 560, 564, 728
Psychology: 270, 456, 504, 505, 519, 685

PART 7: Textbooks Related to Bioassay

Agriculture: 47
Design of Experiments: 422
Elementary Texts: 47, 149, 192, 217, 268, 302, 319, 333, 400, 537, 565, 647, 664, 754, 759
Failure Time: 419
General Texts: 103, 128, 143, 205, 259, 292, 295, 300, 483, 761
Jackknife: 337
Logit Analysis: 48
Medicine: 26, 118, 375
Microbiology: 541
Multivariate: 92, 661, 662
Pesticides: 56
Pharmacology: 674
Probit Analysis: 221, 222, 704
Toxicology: 476, 638

Bibliography

1. Abbott, W. S. 1925. A method of computing the effectiveness of an insecticide. J. of Economic Entomology 18:265–7.
2. Abdelbasit, K. M. and Plackett, R. L. 1981. Experimental design for categorized data. Intern. Statist. Rev. 49:111–26.
3. Abdelbasit, K. M. and Plackett, R. L. 1982. Experimental design for joint action. Biometrics 38:171–9.
4. Abdelbasit, K. M. and Plackett, R. L. 1983. Experimental design for binary data. J. Amer. Stat. Assoc. 78(381):90–8.
5. Addelman, S. 1970. Computing the analysis of variance table for experiments involving qualitative factors and zero amounts of quantitative factors. American Statistician 28(1):21–2.
6. Afsarinejad, K. 1983. Balanced repeated measurements designs. Biometrika 70(1):199–204.
7. Aitchison, J., and Bennett, J. A. 1970. Polychotomous quantal response by maximum indicant. Biometrika 57(2):253–62.
8. Aitchison, J. and Silvey, S. D. 1957. The generalization of probit analysis to the case of multiple responses. Biometrika 44:131–43.
9. Alarie, Y. 1981. Bioassay for evaluating the potency of irritants and predicting acceptable levels of exposure in man. Food and Cosmetics Toxicology 19(5):623–6.
10. Albert, A. and Anderson, J. A. 1981. Probit and logistic discriminant functions. Commun. Statist., A, 10(7):223–7.
11. Ali, A., Carter, E. M. and Hubert, J. J. 1984. Potency estimation in Radioimmunoassay. Tech. Rep., Dept. of Math. and Stat., Univ. of Guelph, Stat. Series No. 1984–163, 12p.
12. Allen, D. M. 1971. The use of prior information in the testing of new drugs. Biometrics 27:699–708.
13. Allen, O. B. 1983. A guide to the analysis of growth curve data with special reference to SAS. Computers and Biomedical Res. 16:101–15.
14. Ameniya, T. 1974. Bivariate probit analysis: minimum chi-square methods. J. Amer. Stat. Assoc. 69(348):940–4.
15. Amemiya, T. 1980. The n^{-2}-order mean squared errors of the maximum likelihood and the minimum logit chi-square estimator. Annals of Statistics 8(3):488–505. (See corrections in (1984) 12(2):783.)
16. Amini, S. B. 1984. A FORTRAN program implemented in a micro computer for estimation of a lethal dose level. Paper for ASA Conf., Aug. 14.
17. Ammann, L. P. 1984. Bayesian nonparametric inference for quantal response data. Annals of Statistics 12(2):636–45.
18. Ammann, L. P. 1984. Large sample properties of Bayesian tolerance curve estimates. Paper for Inst. Math. Stat., Lake Tahoe, Aug. 21, 1984.
19. Anbar, D. 1977. The application of stochastic approximation methods to the bio-assay problem. J. Stat. Planning and Inference 1:191–206.
20. Anderson, J. A. and Philips, P. R. 1981. Regression, discrimination and measurement models for ordered categorical variables. Applied Statistics 30(1):22–31.
21. Anscombe, F. J. 1949. Note on a problem in probit analysis. Annals of Applied Biology 36:203–5.
22. Anscombe, F. J. 1956. On estimating binomial response relations. Biometrika 43:461–4.
23. Aranda-Ordaz, F. J. 1981. On two families of transformations to additivity for binary response data. Biometrika 68(2):357–63.
24. Armitage, P. 1959. Host variability in dilution experiments. Biometrics 15:1–9.
25. Armitage, P. 1970. The combination of assay results. Biometrika 57:665–6.
26. Armitage, P. 1971. Statistical Methods in Medical Research. New York: John Wiley and Sons.
27. Armitage, P. and Allen, I. 1950. Methods of estimating the LD50 in quantal response data. J. of Hygiene 48:298–322.
28. Armitage, P., Bailey, J. M., Petrie, A., Annable, L. and Stack-Dunne, M. P. 1974. Studies in the combination of bioassay results. Biometrics 30:1–9.
29. Armitage, P. and Bennett, B. M. 1974. Maximum likelihood solutions for the combination of relative potencies. J. of Hygiene 73:97–9.
30. Armitage, P., Bennett, B. M. and Finney, D. J. 1976. Point and interval estimation in the combination of bioassay results. J. of Hygiene 76:147–62.
31. Armitage, P. and Spicer, C. C. 1956. The detection of variation in host susceptibility in dilution counting experiments. J. of Hygiene 54:401–14.
32. Arvesen, J. N. and Salsburg, D. S. 1971. Approximate test and confidence intervals using the jackknife. Dept. of Stat. Mimeograph series No. 267, Purdue Univ. (Also appears in "Perspectives in Biometrics" 1: 123–47, R. M. Elashoff (ed.) New York: Academic, 1975.)
33. Asano, C. 1960. A note of modified two-sample-theoretical estimation on biological assay. Bulletin of Mathematical Statistics 9:41–56.
34. Asano, C. 1961. An extensive direct assay in biological assay. Bulletin of Mathematical Statistics 10:1–16.
35. Ashford, J. R. 1958. A problem of subjective classification in industrial medicine. Applied Statistics 8:168–85.

36. Ashford, J. R. 1958. Quantal responses to mixtures of poisons under conditions of simple action—the analysis of uncontrolled data. Biometrika 45:74–88.
37. Ashford, J. R. 1959. An approach to the analysis of data for semi-quantal responses in biological assay. Biometrics 15:573–81.
38. Ashford, J. R. 1981. General models for the joint action of mixtures of drugs. Biometrics 37:457–74.
39. Ashford, J. R., Morgan, D. C., Rae, S. and Sowden, R. R. 1970. Respiratory symptoms in British coal miners. American Review of Respiratory Diseases 102:370–81.
40. Ashford, J. R. and Smith, C. S. 1964. A general system of models for the action of mixture of drugs in biological assay. Biometrika 51:413–28.
41. Ashford, J. R. and Smith, C. S. 1965. An analysis of quantal response data in which the measurement of response is subject to error. Biometrics 21(4):811–25.
42. Ashford, J. R. and Smith, C. S. 1965. An alternative system for the classification of mathematical models for quantal responses to mixtures of drugs in biological assay. Biometrics 21:181–8.
43. Ashford, J. R. and Smith, C. S. 1966. Models for the non-interactive joint action of a mixture of stimuli in biological assay. Biometrika 53:49–60.
44. Ashford, J. R., Smith, C. S. and Brown, S. 1960. The quantal response analysis of a series of biological assays on the same subjects. Biometrika 47:23–32.
45. Ashford, J. R. and Sowden, R. R. 1970. Multivariate probit analysis. Biometrics 26:535–49.
46. Ashford, J. R. and Walker, P. J. 1972. Quantal response analysis for a mixture of populations. Biometrics 28:981–8.
47. Ashton, G. C. and McMillan, I. 1981. A Medley of Statistical Techniques for Researchers. Dubuque: Kendall-Hunt Pub. Co.
48. Ashton, W. D. 1972. The Logit Transformation: with Special Reference to its Use in Bioassay. New York: Hafner.
49. Atkins, E. L. and Anderson, L. D. 1954. Toxicity of pesticide dusts to honeybees. J. of Economic Entomology 47(6):969–72.
50. Ayer, M., Brunk, H. D., Weing, G. M., Reid, W. T. and Silverman, E. 1955. An empirical distribution function for sampling with incomplete information. Annals of Mathematical Statistics 26:641–7.
51. Bailey, N. T. J. 1956. Significance tests for a variable chance of infection in chain-bionomial theory. Biometrika 43:332–6.
52. Bailey, R. C., Homer, L. D., Summe, J. P., McCracken, L. E. and Tobey, R. 1976. Modeling and estimation for time-dependent quantal response. Tech. Rep., Dept. of Clinical and Experimental Immunology, Naval Medical Research Institute, Bethesda, Md.
53. Baksalary, J. K., Corsten, L. C. A. and Kala, R. 1978. Reconciliation of two different views on estimation of growth curve parameters. Biometrika 65(3):662–5.
54. Ballantyne, E. R., Hill, R. K. and Spencer, J. W. 1977. Probit analysis of thermal sensation assessments. Int. J. Biometeorology 21(1):29–43.
55. Ballard, J. A. and Oliff, W. D. 1969. A rapid method for measuring the acute toxicity of dissolved materials to marine fishes. Water Research 3:313–33.
56. Banki, L. 1978. Bioassay of Pesticides in the Laboratory. Budapest: Akadémiai Kiadó.
57. Barr, A. J., Goodnight, J. H., Sall, J. P. and Helwig, J. T. 1976. A User's Guide to SAS • 76. North Carolina: SAS Inst. Inc. (See p. 206.)
58. Barr, M. and Nelson, J. W. 1949. An accurate and economical method for the biological assay of aconite tincture. J. Amer. Pharm. Assoc. 38:518–21.
59. Bartlett, M. S. 1937. Subsampling for attributes. J. Roy. Stat. Soc. Supp. 4:131–5.
60. Bartlett, M. S. 1937. Some examples of statistical methods of research in agriculture and applied biology. J. Roy. Stat. Soc. Supp. 4:137–70.
61. Bartlett, M. S. 1946. A modified probit technique for small probabilities. J. Roy. Stat. Soc. Supp. 8:113–7.
62. Bartlett, M. S. 1947. The use of transformations. Biometrics 3:39–52.
63. Basu, D., and Fagerstrom, R. A. 1979. Bayesian approach to bioassay. Tech. Rep. No. M398, Dept. of Stat., Florida State Univ.
64. Bazhanov, V. S. 1971. Method of probit assay in comparative evaluation of rubomycin efficiency in the treatment of transplantable tumors in mice. Antibiotiki (Moscow) 16(8):714–8.
65. Behrens, B. 1929. Zur Auswertung der Digitalisblatter im Froschversuch. Arch. Exp. Path. Pharmak. 140:237–56.
66. Bell, P. M. 1973. A programming device for bioassay. J. of Physiology 234(2):7–8.
67. Bennett, B. M. 1952. Estimation of LD50 by moving averages. J. of Hygiene 50:157–64.
68. Bennett, B. M. 1959. On a multivariate version of Fieller's theorem. J. roy. Stat. Soc., B, 21:59–62.
69. Bennett, B. M. 1961. Confidence limits for multivariate ratios. J. Roy. Stat. Soc., Ser., B, 23:108–12.
70. Bennett, B. M. 1962. On combining estimates of relative potency in bioassay. J. of Hygiene 60:378–86.
71. Bennett, B. M. 1963. Slope ratio assays and confidence limits. Metrika 7:117–20.
72. Bennett, B. M. 1963. On combining estimates of a ratio of means. J. Roy. Stat. Soc., A, 126:201–5.
73. Bennett, B. M. 1964. A note on combining correlated estimates of a ratio of multivariate means. Technometrics 6:463–7.

74. Bennett, B. M. 1978. On the treatment of binary data by means of ranks. Biometrical Journal 20(7):749–54.
75. Bennett, B. M. 1984. On the use of ranks in combined estimates in quantal bioassays. Biometrical Journal 26:713.
76. Berkson, J. 1944. Application of the logistic function to bioassay. J. Amer. Stat. Assoc. 39:357–65.
77. Berkson, J. 1946. Approximation of chi-square by 'probits' and 'logits'. J. Amer. Stat. Assoc. 41:70–4.
78. Berkson, J. 1949. Minimum X^2 and maximum likelihood solutions in terms of a linear transform, with particular reference to bio-assay. J. Amer. Stat. Assoc. 44:273–8.
79. Berkson, J. 1950. Are there two regressions? J. Amer. Stat. Assoc. 45:164–80.
80. Berkson, J. 1950. Some observations with respect to the error of bioassay. Biometrics 6:432–4.
81. Berkson, J. 1951. Why I prefer logits to probits. Biometrics 7:327–39.
82. Berkson, J. 1953. A statistically precise and relatively simple method of estimating the bioassay with quantal response based on the logistic function. J. Amer. Stat. Assoc. 48:565–99.
83. Berkson, J. 1955. Maximum likelihood and minimum x^2 estimates of the logistic function. J. Amer. Stat. Assoc. 50:130–62.
84. Berkson, J. 1955. Estimate of the integrated normal curve by minimum normit chi-square with particular reference to bio-assay. J. Amer. Stat. Assoc. 50:529–50.
85. Berkson, J. 1957. Tables for the use in estimating the normal distribution function of normit analysis. Biometrika 44:411–53.
86. Berkson, J. 1960. Nonograms for fitting the logistic function by maximum likelihood. Biometrika 47:121–42.
87. Berkson, J. 1980. Minimum chi-square, not maximum likelihood (with discussion). Annals of Statistics 8(3):457–87.
88. Bhan, A. K. 1974. Computing LD50 more efficiently and accurately: a modification of the Reed–Muench method. Newsletter–Indian Soc. for Nuclear Tech. in Agr. and Biology 3(1):5–6.
89. Bhattacharya, P. K. 1981. Posterior distribution of a Dirichlet process from quantal response data. Annals of Statistics 9(4):803–11.
90. Bickel, P. J. and Doksum, J. A. 1981. An analysis of transformations revisited. J. Amer. Stat. Assoc. 76:296–311.
91. Biggers, J. D. 1952. The calculation of the dose-response line in quantal assay with special reference to oestrogen assays by the Allen–Doisy technique. J. Endoc. 8:169–78.
92. Bishop, Y. M. M., Fienberg, S. E. and Holland, P. W. 1975. Discrete Multivariate Analysis: Theory and Practice. Cambridge: MIT Press. (See Ch. 10; also see review by S. J. Haberman, Annals of Statistics (1976) 4:817–20.)
93. Bissell, A. F. and Ferguson, R. A. 1975. The jackknife- toy, tool or two-edged weapon? The Statistician 24:79–100.
94. Bliss, C. I. 1934. The method of probits. Science 79(2037):38–9. (Corrections on p. 409–10)
95. Bliss, C. I. 1935. The calculation of dosage-mortality curve. Annals of Applied Biology 22:134–67.
96. Bliss, C. I. 1935. The comparison of dosage-mortality data. Annals of Applied Biology 22:307–33.
97. Bliss, C. I. 1937. The calculation of the time-mortality curve. Annals of Applied Biology 24:815–52.
98. Bliss, C. I. 1938. The determination of the dosage-mortality curve from small numbers. Quart. J. Pharm. and Pharmacol. 11:192–216.
99. Bliss, C. I. 1939. The toxicity of poisons applied jointly. Annals of Applied Biology 26:585–615.
100. Bliss, C. I. 1939. Fly spray testing: A discussion on the theory of evaluating liquid household insecticides by the Peet-Grady method. Soap 15(4):103–11.
101. Bliss, C. I. 1940. The relation between exposure time, concentration and toxicity in experiments on insecticides. Annals Entom. Soc. Amer. 33:721–66.
102. Bliss, C. I. 1941. Biometry in the service of biological assay. Indust. Eng. Chem. 13:84–8.
103. Bliss, C.I. 1952. The Statistics of Bioassay. New York: Academic.
104. Bliss, C. I. 1957. Bioassay from a parabola. Biometrics 13:35–50.
105. Bliss, C. I. 1957. Some principles in bioassay. Amer. Science 45:449–66.
106. Bliss, C. I. 1964. Fisher's contributions to medicine and bioassay. Biometrics 20:273–85.
107. Bliss, C. I. 1970. Dose-response curves for radioimmunoassays. In: Statistics in Endocrinology (J. W. McArthur and T. Colton, eds.) Cambridge: MIT Press.
108. Bliss, C. I. and Cattell, M. 1943. Biological assay. Ann. Rev. Physiol. 5:479–539.
109. Bliss, C. I. and Griminger, J. 1969. Response criteria for the bioassay of vitamin K. Biometrics 25(4):735–45.
110. Bliss, C. I. and Hanson, J. C. 1939. Quantitative estimation of the potency of digitalis by the cat method in relation to secular variation. J. Amer. Pharm. Assoc. 28:521–30.
111. Bliss, C. I. and Marks, H. P. 1939. The biological assay of insulin. Quart. J. Pharm. Pharmacol. 12:82–110 and 182–205.
112. Boetsch, G. and Bley, D. 1980. Age at menarche in a sample of girls living in Paris France. Bull. Mem. Soc. Anthropol. 7:3–6.
113. Bonnier, G. and Luning, K. G. 1949. Studies on x-ray mutations in the white and forked loci of *Drosophila melanogaster*. A statistical analysis of mutation frequencies. Herediatas 35:163–89.
114. Booth, G. D. 1975. On the use of Abbott's formula: a correction for natural response. Biometrics 31(2):590.

115. Borgmann, U. 1981. Determination of free metal ion concentration using bioassays. Can. J. Fish. Aquat. Sci. 38(8):999–1002.
116. Bowden, D. C. and Steinhorst, R. K. 1973. Tolerance bands for growth curves. Biometrics 29:361–71.
117. Box, G. E. P. and Hay, W. A. 1953. A statistical design for the efficient removal of trends occurring in a comparative experiment with an application in biological assay. Biometrics 9:301–19.
118. Boyd, W. C. 1956. Fundamentals of Immunology, 3rd ed. London: Interscience.
119. Breslow, N. 1981. Odds ratio estimators when the data are sparse. Biometrika 68(1):73–84.
120. Brett, J. R. 1952. Temperature tolerance in young Pacific salmon, genus *Oncorhynchus*. J. Fish. Res. Board Can. 9(6):265–323.
121. Brillinger, D. R. 1964. The asymptotic behavior of Tukey's general method of setting approximate confidence limits (the jackknife) when applied to maximum likelihood estimates. Review of the International Statistical Institute 32(3):202–6.
122. Brindley, W. A. 1975. Insecticide bioassays with field populations of alfalfa weevils, a simplified approach. J. of Economic Entomology 68(2):227–30.
123. Broderius, S. J. and Smith, L. L. 1979. Lethal and sublethal effect of binary mixtures of cyanide and hexavalent chromium, zinc, or ammonia to the fathead minnow (*Pimephales promelas*) and rainbow trout (*Salmo gairdneri*). J. Fish. Res. Board of Can. 36:164–72.
124. Brookmeyer, R. and Crowley, J. 1982. A confidence interval for the median survival time. Biometrics 38:29–42.
125. Bross, I. D. J. 1950. Estimates of the LD50: a critique. Biometrics 6:641–7.
126. Brown, B. W. 1959. Some properties of the Spearman estimator in bioassay. Ph.D. Thesis, Univ. of Minnesota.
127. Brown, B. W. 1961. Some properties of the Spearman estimator in bioassay. Biometrika 48:293–302.
128. Brown, B. W. 1964. Statistical procedures in biological assay. Lecture Notes, Stanford Univ.
129. Brown, B. W. 1966. Planning a quantal assay of potency. Biometrika 22:322–9.
130. Brown, B. W. 1970. Quantal-response assays. In: Statistics in Endocrinology. J. W. McArthur and T. Colton (eds.) Cambridge: MIT Press. (See p. 129–144.)
131. Brown, B. W. 1978. Quantal response. In the International Encyclopedia of Statistics, W. H. Kruskal and J. M. Tanur, editors. Macmillan Pub. Co., 2:816–24.
132. Brown, C. C. 1978. The statistical analysis of dose-effect relationships. In: Principles of Ecotoxicology: Scope. (G. C. Butler, ed.) Ch. 12: 115–148. New York: Wiley.
133. Brown, C. C. 1982. On a goodness of fit test for the logistic model based on score statistics. Communications in Statistics: Theory and Methods 11:1087–1105.
134. Brown, P. J. 1982. Multivariate calibration (with discussion). J. Roy. Stat. Soc., B, 44(3):287–321.
135. Brown, V. M. 1973. Concepts and outlook in testing the toxicity of substances in fish. In: Bioassay Techniques and Environmental Chemistry. (G. F. Glass, ed.) Ann Arbor: Ann Arbor Science Publishers. (See pages 73–95.)
136. Brown, V. M., Jordan, D. H. M. and Tiller, B. A. 1969. The acute toxicity of rainbow trout of fluctuating concentrations and mixtures of ammonia, phenol and zinc. J. Fish Biology 1:1–9.
137. Brown, W. F. 1964. Variance estimation in the Reed-Muench fifty percent end-point determination. Amer. J. Hygiene 79:37–46.
138. Brownlee, K. I., Hodges, J. L. and Rosenblatt, M. 1953. The up-and-down method with small samples. J. Amer. Stat. Assoc. 48:262–77.
139. Brunk, H. D. 1958. On the estimation of parameters restricted by inequalities. Annals of Mathematical Statistics 29:437–54.
140. Bucher, G. E. and Morse, P. M. 1963. Precision of estimates of the median lethal dose of insect pathogens. J. Insect Pathology 5:289–308.
141. Buffler, P. A., Suarez, L., Osborne, R., Carvajal, H. and Larsen, D. 1978. Mortality in a pediatric burns unit: probit analysis. Amer. J. of Epidemiology (Abstract) 108(3):231.
142. Burdick, G. E. 1957. A graphical method of deriving threshold values of toxicity and the equation of the toxicity curve. J. New York Fish and Game Soc. 4(1):102–8.
143. Burn, J. H., Finney, D. J. and Goodwin, L. G. 1950. Biological Standardization, 2nd ed. London: Oxford Univ. Press.
144. Busvine, J. R. 1938. The toxicity of ethylene oxide to *Calandra oryzae, C. granaria, Tribolium castaneum* and *Cimex lectularius*. Annals of Applied Biology 25:605–32.
145. Busvine, J. R. 1957. A Critical Review of the Technique for Testing Insecticides. London: Commonwealth Institute of Entomology.
146. Campbell, F. L. 1930. A comparison of four methods for estimating the relative toxicity of stomach poison insecticides. J. of Economic Entomology 23:357–70.
147. Campbell, F. L. and Moulton, F. R. 1943. Laboratory Procedures in Studies of the Chemical Control of Insects. Washington: AAAS.
148. Campbell, R. C. 1966. The chick assay of lysine. Biometrics 22(1):58–73.
149. Campbell, R. C. 1974. Statistics for Biologist, 2nd ed. London: Cambridge Univ. Press.

150. Carpenter, K. J., March, B. E., Milner, C. K. and Campbell, R. C. 1963. A growth assay with chicks for the lysine content of protein concentrates. British J. of Nutrition 17:309–23.
151. Carter, E. M. and Hubert, J. J. 1981. Growth curve models for multivariate quantal bioassays. Tech. Rep. No. 1981–129, Dept. of Math. and Stat., Univ. of Guelph, 33p.
152. Carter, E. M. and Hubert, J. J. 1981. Covariable adjustment for multivariate growth curve models in quantal bioassays. Tech. Rep. No. 1981–130, Dept. of Math. and Stat., Univ. of Guelph, 14p.
153. Carter, E. M. and Hubert, J. J. 1981. Confidence limits and confidence bands for parameters based on multivariate models in bioassay. Tech. Rep. No. 1981–131, Dept. of Math. and Stat., Univ. of Guelph, 22p.
154. Carter, E. M. and Hubert, J. J. 1983. A test for constant potency in multivariate bioassay. Tech. Rep. No. 1983–151, Dept. of Math. and Stat., Univ. of Guelph, 22p.
155. Carter, E. M. and Hubert, J. J. 1984. A growth-curve model approach to multivariate quantal bioassay. Biometrics 40(3):699–706.
156. Carter, E. M. and Hubert, J. J. 1984. Covariable adjustment in multivariate bioassay. ASA Conference Proceedings 1984:103–6.
157. Carter, E. M. and Hubert, J. J. 1985. Analysis of parallel-line assays with multivariate responses. Biometrics 41(3):703–710.
158. Carter, E. M. and Hubert, J. J. 1985. The asymptotic distribution of the maximum likelihood estimator of the relative potency. Tech. Rep. No. 1983–158, Dept. of Math. and Stat., Univ. of Guelph, 19p. (Submitted to Canadian J. of Statistics)
159. Carter, E. M., Walsh, M. N. and Hubert, J. J. 1985. Analysis of multivariate parabolic bioassays. Submitted to J. Roy. Stat. Soc. B.
160. Cernosek, S. F. Jr., Quint, J. and Stirtz, D. 1980. Evaluation of the goodness-of-fit of radio immunoassay data using various curve fitting methods. Clinical Chemistry 26(7):1007–8.
161. Chakravarti, N. K. 1960. The mathematical theory of biological assay of a local anaesthetic. Biometrics 16:278–91.
162. Chambers, E. A. and Cox, D. R. 1967. Discrimination between alternative binary response models. Biometrika 54:573–8.
163. Chand, N. and Hoel, W. 1974. A comparison of models for determining safe levels of environmental agents. In: Reliability and Biometry (p. 681). F. Proschan and R. J. Serfling, eds. Philadelphia: SIAM.
164. Chang, P. C. 1973. Existence of a constant relative potency in bio-assays. Preprint, Div. of Biostatistics, School of Public Health, UCLA.
165. Chang, P. C. 1974. Estimation of percentiles from quantal data with special attention to planning. Preprint, Div. of Biostatistics, School of Public Health, UCLA.
166. Chang, P. C. and Johnson, E. A. 1972. Some distributions-free properties of the asymptotic variance of the Spearman estimator in bioassays. Biometrics 28(3):882–9.
167. Chapman, D. I. 1979. Radioimmunoassay. Chemistry in Britain 15:439–47.
168. Chen, C. W. 1969. A kinetic model of fish toxicity threshold. Ph.D. Thesis, Univ. of California, Berkeley.
169. Chen, C. W. and Selleck, R. E. 1969. A kinetic model of fish toxicity threshold. J. Water Pollution Control Federation 41:R295–308.
170. Chen, K-H. 1980. Moment analysis for the probit model of directional selection. Biometrical Journal 21(8):773–80.
171. Chen, K-H. and Price, R. A. 1980. Familial variance covariance structure under probit mate selection. Behavior Genetics 10(5):471–2.
172. Chew, R. D. 1982. Estimating toxicity curves by fitting a compartment-based model to median survival times. Ph.D. Thesis, Montana St. Univ.
173. Chew, R. D. and Hamilton, M. A. 1985. Toxicity curve estimation: fitting a compartment model to median survival times. (To appear in J. Amer. Fisheries Soc.)
174. Chhikara, R. S. and Guttman, I. 1982. Prediction limits for the inverse Gaussian distribution. Technometrics 24(4):319–24.
175. Chikuse, Y. 1980. Pseudo confidence regions for the solution of a multivariate linear functional relationship. Tech. Rep. No. 80–16, Dept. of Stat., Univ. of Pittsburgh.
176. Chmiel, J. J. 1976. Some properties of Spearman-type estimators of the variance and percentiles in bioassay. Biometrika 63(3):621–6.
177. Choi, S. C. 1971. An investigation of Wetherill's method of estimation for the up-and-down experiment. Biometrika 27:961–70.
178. Christensen, E. R. 1984. Dose-response functions in aquatic testing and the Weibull model. Water Research 18:213–221.
179. Christensen. E. R. and Chen, C. Y. 1985. A general noninteractive multiple toxicity model including probit, logit, and Weibull transformations. Biometrics 41(3):711–25.
180. Church, J. D. and Cobb, E. B. 1971. Non-parametric estimation of the mean using quantal response data. Ann. Inst. Statist. Math. 23(1):105–17.
181. Church, J. D. and Cobb, E. B. 1973. On the equivalence of Spearman-Kärber and maximum likelihood of the mean. J. Amer. Stat. Assoc. 68:201–2.

182. Church, J. D. and Cobb, E. B. 1976. Statistical methods for small-sample bioassays. Contributed paper, 9th Inter. Biometric Conf., Boston.
183. Claringbold, P. J. 1956. The within animal bioassay with quantal responses. J. Roy. Stat. Soc., B, 18:133–7.
184. Claringbold, P. J. 1959. Orthogonal contrasts in slope ratio investigations. Biometrics 15:307–22.
185. Claringbold, P. J., Biggers, J. D. and Emmens, C. W. 1953. The angular transformation in quantal analysis. Biometrics 9:467–84.
186. Cobb, E. B. and Church, J. D. 1983. Small-sample quantal response methods for estimating the location parameter for a location-scale family of dose-response curves. J. Amer. Stat. Assoc. 78(381):99–107.
187. Cochran, W. G. 1973. Experiments for nonlinear functions. J. Amer. Statist. Assoc. 68:771–81.
188. Cochran, W. G. and Davis, M. 1965. The Robbins-Monro method for estimating the median lethal dose. J. Roy. Stat. Soc., B, 27(1):28–44.
189. Cochran, W. G. and Davis, M. 1961. Sequential experiments for estimation the median lethal dose. Colloques Internationaux du Centre National de la Récherche Scientifique No. 110, Paris.
190. Cochran, W. G. and Davis, M. 1963. Sequential experiments for estimating the median lethal dose. Le Plan d'Expériences, 181–94, Paris. Centre Nationale de la Récherche Scientifique.
191. Cochran, W. G. and Davis, M. 1964. Stochastic approximation to the median effective dose in bioassay Stochastic Models in Medicine and Biology. J. Gurland (ed.) Madison: Univ. of Wisconsin Press (p. 281–99).
192. Colquhoun, D. 1971. Lectures on Biostatistics. London: Oxford Un. Press.
193. Cook, B. 1974. Automation and data processing for radioimmunoassays. Steroid Immunoassay, Proceedings of the 5th Tenovus Workshop.
194. Copenhaver, T. W. and Mielke, P. W. 1977. Quantit analysis: a quantal assay refinement. Biometrics 33(1):175–86.
195. Copenhaver, T. W. and Mielke, P. W. 1977. Quantit analysis: a generalized tolerance distribution for quantal response assay analysis. Paper presented at the Joint Stat. Conf., Chicago, Aug. 16, 1977.
196. Cornell, R. G. and Petersen, V. J. 1970. Evaluation of a quantal response model with estimated concentration. Biometrics 26(4):713–22.
197. Cornfield, J. 1954. Measurement and comparison of toxicities: the quantal response. Statistics and Mathematics in Biology, O. Kempthorne et. al. (eds.) Ames: Iowa State College Press.
198. Cornfield, J. 1964. Comparative bioassays and the role of parallelism. J. of Pharmacology and Experimental Therapeutics 144:143–9.
199. Cornfield, J. 1967. The meaning of bioassay: a comment. (Letter) Biometrics 23:160–2.
200. Cornfield, J. and Mantel, N. 1950. Some new aspects of the application of maximum likelihood to the calculation of the dosage response curve. J. Amer. Stat. Assoc. 45:181–210.
201. Cornfield, J., Rai, K. and Van Ryzin, J. 1980. Procedures for assessing risk at low levels of exposure. Arch. Toxicol. 3 (Suppl.):295.
202. Cox, C. 1981. Quantal response bioassay with extra-binomial variables. Tech. Rep. No. 81–13, Dept. of Stat., Univ. of Rochester, 28p.
203. Cox, C. P. 1972. On estimating relative potency from quadratic log-dose response relationships. Biometrics 28(3):875–81.
204. Cox, C. P. and Leaverton, P. E. 1966. Statistical procedures for bioassays when the condition of similarity is not satisfied. J. of Pharmacological Science 55:716–23.
205. Cox, D. R. 1970. Analysis of Binary Data. London: Methuen.
206. Cramer, E. M. 1962. A comparison of three methods of fitting the normal ogive. Psychometrika 27:183–92.
207. Cramer, E. M. 1964. Some comparisons of methods of fitting the dosage response curve for small samples. J. Amer. Stat. Assoc. 59:779–93.
208. Cranmer, M. F. 1974. Reflections in toxicology. J. of the Washington Academy of Sciences 64(2):158–79.
209. Creasy, M. A. 1954. Limits for the ratio of means. J. Roy. Stat. Soc., A, 117:186–94.
210. Creasy, M. A. 1956. Confidence limits for the gradient in the linear functional relationship. J. Roy. Stat. Soc., B, 18:65–9.
211. Cress, P., Dirkson, P. and Graham, J. W. 1968. Fortran IV and Watfor. Englewoods Cliffs: Prentice-Hall.
212. Crump, K. S. 1977. Low dose extrapolation utilizing the age distribution of cancer. Paper presented at the 137th Annual ASA Meeting, Chicago.
213. Crump, K. S., Guess, H. A. and Deal, K. L. 1977. Confidence intervals and tests of hypotheses concerning dose-response relations inferred from animal carcinogenicity data. Biometrics 33(3):437–51.
214. Daffer, P. Z., Crump, K. S. and Masterman M. D. 1980. Asymptotic theory for analyzing dose response survival data with application to the low-dose extrapolation problem. Math. Biosci. 50:207–30.
215. Dagenais, D. L. 1983. The use of a probit model for the validation of selection procedures. Paper for Can. Stat. Soc. Conf., Ottawa, June.
216. Dagenais, D. L. 1983. Estimation of the probit model with serial correlation. Paper for Can. Stat. Soc. Conf., Ottawa, June.
217. Daniel, W. W. 1974. Biostatistics: A Foundation for Analysis in the Health Sciences. Toronto: Wiley.
218. Darby, S. C. 1980. A Bayesian approach to parallel line bioassay. Biometrika 67(3):607–12.
219. Das, M. N. and Kulkarni, G. A. 1966. Incomplete block designs for bioassays. Biometrics 22:706–29.

220. Daum, R. J. 1970. A revision of two computer programs for probit analysis. Bull. Ent. Soc. Amer. 16:10–5.
221. Daum, R. J. and Givens, C. 1963. Potency Probit Analysis. Beltsville: Biometrical Services, USDA.
222. Daum, R. J., Givens, C. and Bearden, G. 1962. Probit Analysis. Beltsville: Biometrical Services, USDA.
223. David, E. E. 1971. Ecological effects of pesticides on non-target species. Special Report, Office of Science and Technology, U.S. Government Printing Office.
224. David, L., and Jayewardene, C. H. 1974. Age at menarche in Ceylon. Ann. Human Biology 1(4):447–9.
225. Davis, J. C. and Hoos, R. A. W. 1975. Use of sodium pentachlorophenate dehydroabietic acid as reference toxicants for salmonoids bioassays. J. Fish. Res. Board Can. 32(3):411–6.
226. Davis, L. 1984. Comments on a paper by T. Amenmiya on estimation in a dichotomous logit regression model. Annals of Statistics 12(2):778–82.
227. Davis, M. 1965. Sequential experiments in bioassay. Ph.D. Thesis, Harvard Un.
228. Davis, M. 1971. Comparison of sequential bioassays in small samples. J. Roy. Stat. Soc., B, 33(1):78–87.
229. Davis, O. L. 1980. Note on regression with correlated responses (letter). Biometrics 36(3):551–2.
230. Davis, O. L. and Budgett, D. A. 1980. Accelerated storage tests on pharmaceutical products: effect of error structure of assay and errors in recorded temperature. J. of Pharmacy and Pharmacology 32:155–9.
231. Davis, R. G. 1971. Computer Programming in Quantitative Biology. New York: Academic Press.
232. DeBeer, E. J. 1941. A scale for graphically determining the slopes of dose-response curves. Science 94:521–2.
233. DeBeer, E. J. 1945. The calculation of biological assay results by graphical methods. The all-or-none type of response. J. of Pharmacology and Experimental Therapeutics 85:1–13.
234. DeLean, A., Munson, P. J. and Rodbard, D. 1978. Simultaneous analysis of families of sigmoidal curves. Amer. J. of Physiology 235(2):97–102.
235. De March, B. G. E. 1983. Models for the joint effects of toxicants in acute lethal bioassays. Technical report, presented at the Annual Toxicity Workshop, Halifax, N.S., Nov. 7–10.
236. Dempster, A. P., Laird, N. M. and Rubin, D. M. 1977. Maximum likelihood from incomplete data via the EM algorithm. J. Roy. Stat. Soc., B, 38:1–11.
237. Diaconis, P. and Efron, B. 1983. Computer-intensive methods in statistics. Scientific American, May: 116–30.
238. Dick, J. P., Carter, E. M. and Hubert, J. J. 1983. A multivariate analysis of time dependent parallel-line assays. Paper for the Joint Stat. Conf., Toronto, Aug., 22p.
239. Dietrich, F. H. and Marks, R. G. 1978. Analysis of a factorial quantal response assay using inverse regression. Paper for the Joint Stat. Conf., San Diego, Aug., 1978.
240. Disch, D. 1981. Bayesian nonparametric inference for effective doses in a quantal-response experiment. Biometrics 37(4):713–22.
241. Dixon, W. J. 1965. The up-and-down method for small samples. J. Amer. Stat. Assoc. 60:967–78.
242. Dixon, W. J. and Brown, M. B. (ed.) 1979. BMDP–79. Berkeley: Univ. of California Press.
243. Dixon, W. J. and Mood, A. M. 1948. A method for obtaining and analyzing sensitivity data. J. Amer. Stat. Assoc. 43:109–26.
244. Dixon, W. J. and Mood, A. M. 1968. The up-and-down method for small samples. J. Amer. Stat. Assoc. 60:961–70.
245. Donnelly, Y. G. M., McKean, H. E., Heird, C. S. and Green, J. 1981. Bioassay of a cigarette smoke fraction. 2. Experimental design and potency estimations. J. of Toxicology and Environmental Health 7(3–4):419–44.
246. Dorfman, D. and Cirello, J. 1978. A bioassay to determine subacute effects of toxicants. (Abstract) Bulletin, New Jersey Academy of Science 3(2):93.
247. Doudoroff, P., Anderson, B. G., Burdick, G. E., Galtsoff, P. S. and Hart, W. B. 1951. Bio-assay methods for the evaluation of acute toxicity of industrial wastes to fish. Sewage and Industrial Wastes 23(110):1380–97.
248. Doudoroff, P. and Katz, M. 1953. Critical review of literature on the toxicity of industrial wastes and their components to fish. II. The metals, as salts. Sewage Industrial Wastes (Now renamed J. of Water Pollution Control) 25(7):802–39.
249. Dragstedt, C. A. and Lang, V. F. 1928. Respiratory stimulants in acute poisoning in rabbits. J. of Pharmacology 32:215–22.
250. Drane, J. W., Owen, D. B. and Serbert, G. B. 1977. The Burr distribution and quantal bioassay. Paper for the Biometric Society Meeting, Univ. of North Carolina, April 18, 1977. (See abstract in Biometrics 33(3):578.)
251. Ducros, A. 1980. Age at menarche in Easter Island Pacific Ocean. Bull. Mem. Soc. Anthropol. 7:7–12.
252. Dunachie, J. F. and Fletcher, W. W. 1969. An investigation of the toxicity of insecticides to birds' eggs using the egg-injection technique. Annals of Applied Biology 64:409–23.
253. Dunsmore, I. R. 1981. Growth curves in two-period change over models. Applied Statistics 30(3):223–9.
254. Durling, F. C. 1974. Bivariate probit, logit and burrit analysis. Biometrics 30:378 (Abstract).
255. Eckhart, A. C. and Mahoney, B. L. Jr. 1971. A bioassay study of the effects of copper and zinc in the Rapp Ahannock River (Abstract). Virginia J. of Science 22(3):111.

256. Eichhorn, B. H. 1974. Sequential search of an optimal dosage for cases of linear dosage—toxicity regression. Commun. in Stat. 3(3):263–71.
257. Eisenhart, C. 1939. The interpretation of certain regression methods and their use in biological and industrial research. Annals of Mathematical Statistics 10:162–86.
258. Elashoff, J. D. 1981. Repeated-measures bioassay with correlated errors and heterogeneous variances: a Monte Carlo Study. Biometrics 37:475–82.
259. Elashoff, R. M. (ed.) 1975. Perspectives in Biometrics, Vol 1. New York: Academic Press.
260. Eloranta, V. 1975. Modified bioassay procedure for toxic effluents. J. of Water Pollution Control 47(8):2172–7.
261. Elston, R. C. 1965. A simple method of estimating relative potency from two parabolas. Biometrics 21:140–9.
262. Elston, R. C. 1969. An analogue to Fieller's Theorem using Scheffe's solution to the Fisher-Behrens problem. Amer. Statist. 23(1):26–8.
263. Elston, R. C. and Grizzle, J. E. 1962. Estimation of time-response curves and their confidence bands. Biometrics 18:148–59.
264. Environmental Protection Agency. 1971. Effects of chemicals on aquatic life. Water Quality Criteria Data Book, Vol 3 (Selected data from the literature through 1968). See sections 4 and 7.
265. Epstein B. and Churchman, C. W. 1944. On the statistics of sensitivity data. Annals of Mathematical Statistics 15:90–6.
266. Falkoff, A. D. and Iverson, K. E. 1968. APL/360: A User's Manual. White Plains: IBM Corporation.
267. Feder, P. I. and Sherrill, M. E. 1980. A computer program to calculate nonparametric lower confidence bounds on safe concentrations in quantal response toxicity tests. Tech. Rep. No. 215, Dept. of Stat., Ohio State Univ., 25p.
268. Federer, W. T. 1973. Statistics and Society. New York: Marcel Dekker.
269. Feistel, R. F. 1972. The jackknife statistic and its use in setting approximate confidence intervals. Master's Thesis, S. F. Austin State Univ.
270. Ferguson, G. A. 1942. Item selection by the constant process. Psychometrika 7:19–29.
271. Fieller, E. C. 1932. The distribution of the index in a normal bivariate population. Biometrika 24:428–40.
272. Fieller, E. C. 1940. The biological standardization of insulin. J. Roy. Stat. Soc. Supp. 7:1–64.
273. Fieller, E. C. 1944. A fundamental formula in the statistics of biological assay and some applications. Quart. J. Pharm. and Pharmacol. 17:117–23.
274. Fieller, E. C. 1947. Some remarks on the statistical background in bio-assay. Analyst 72:37–43.
275. Fieller, E. C. 1954. Some problems in interval estimation. J. Roy. Stat. Soc., B, 16:175–85.
276. Fieller, E. C., Creasy, M. A. and David, S. T. 1954. Symposium on interval estimation. J. Roy. Stat. Soc., B, 16:175–222.
277. Fieller, E. C., Irwin, J. O., Marks, H. P. and Shrimpton, E. A. G. 1939. The dosage-response relation in the cross-over rabbit test for insulin. Quart. J. Pharm. and Pharmacol. 12:206–11 and 724–42.
278. Finkel, D. J. 1976. Probit analysis of upper pleistocene hominids. Amer. J. Physical Anthropology 44(1):177–8.
279. Finney, D. J. 1938. The distribution of the ratio estimates of the two variances in a sample from a normal biavariate population. Biometrika 30:190–2.
280. Finney, D. J. 1942. The analysis of toxicity tests on mixtures of poisons. Annals of Applied Biology 29:82–94.
281. Finney, D. J. 1943. The design and interpretation of bee experiments. Annals of Applied Biology 30:197.
282. Finney, D. J. 1944. The application of the probit method to toxicity test data adjusted for mortality in the controls. Annals of Applied Biology 31:68–74.
283. Finney, D. J. 1945. The microbiological assay of vitamins: the estimate and its precision. Quart. J. Pharm. and Pharmacol. 18:77–82.
284. Finney, D. J. 1946. The analysis of a factorial series of insecticide tests. Annals of Applied Biology 33:160–5.
285. Finney, D. J. 1946. The design and statistical analysis of microbiological assays. Quart. J. Pharm. and Pharmacol. 19:112–7.
286. Finney, D. J. 1947. The principles of biological assay. J. Roy. Stat. Soc. Supp. 9:46–91.
287. Finney, D. J. 1947. The estimation from individual records of the relationship between dose and quantal response. Biometrika 34:320–34.
288. Finney, D. J. 1949. The adjustment for a natural response rate in probit analysis. Annals of Applied Biology 36:187–95.
289. Finney, D. J. 1950. The estimation of the mean of a normal tolerance distribution. Sankhya 10:341–60.
290. Finney, D. J. 1951. Subjective judgment in statistical analysis: an experimental study. J. Roy. Stat. Soc., B, 13:284–97.
291. Finney, D. J. 1952. The estimation of the ED50 for a logistic response curve. Sankhya 12:121–36.
292. Finney, D. J. 1955. Experimental Design. Chicago: Univ. of Chicago Press. London: C. Griffin.
293. Finney, D. J. 1965. The meaning of bioassay. Biometrics 21:785–98.
294. Finney, D. J. 1970. Covariance analysis in bioassay and related problems. In: Statistics in Endocrinology (J. W. McArthur and T. Colton, eds.). (See pages 163–192.)
295. Finney, D. J. 1971. Probit Analysis, 3rd ed. New York: Cambridge U. Press.

296. Finney, D. J. 1976. Estimation of the response curve in radiological assays. Contributed paper, Biometrics Conference, Boston.
297. Finney, D. J. 1976. A computer program for parallel line bioassays. J. of Pharmacology and Experimental Therapeutics 198:497–506.
298. Finney, D. J. 1976. Radioligand assay. Biometrics 32:721–40.
299. Finney, D. J. 1977. The statistical structure of biological assays. J. of Japanese Stat. Soc. 7:1–9.
300. Finney, D. J. 1978. Statistical Method in Biological Assay, 3rd ed. Griffin: London.
301. Finney, D. J. 1979. Bioassay and the practice of statistical inference. Intern. Stat. Rev. 47:1–12.
302. Finney, D. J. 1980. Statistics for Biologists. New York: Chapman–Hall.
303. Finney, D. J., Holt, L. B. and Sheffield, F. 1975. Repeated estimations of an immunological response curve. J. of Biol. Standard. 3:1–10.
304. Finney, D. J. and Phillips, P. 1977. The form and estimation of a variance function, with particular reference to radioimmunoassay. Applied Statistics 26:312–20.
305. Finney, D. J. and Schild, H. O. 1966. Parallel line assay with successive adjustment of doses. Brit. J. of Pharm. and Chemotherapy 28:84–92.
306. Finney, D. J. and Stevens, W. L. 1948. A table for the calculation of working probits and weights in probit analysis. Biometrika 35:191–201.
307. Fisher, R. A. 1935. Appendix to Bliss, C. I.: The case of zero survivors. Annals of Applied Biology 22:164–6.
308. Fisher, R. A. 1949. A biological assay of tuberculin. Biometrics 5:300–16.
309. Fleming, W. E. and Baker, F. E. 1932. Hot water as an insecticide for the Japanese beetle in soil and its effect on the roots of nursery plants. Tech. Bull. #274, U.S. Dept. of Agr.
310. Forsey, D. R., Carter, E. M. and Hubert, J. J. 1981. MASSAY: An APL program for multivariate quantal bioassays. Tech. Rep. No. 1981–132, Dept. of Math. and Stat., Univ. of Guelph, 32p.
311. Fraser, D. A. S. 1979. Inference and Linear Models. McGraw-Hill: Toronto. (See Chapter 10.)
312. Fraser, D. A. S. and Prentice, R. L. 1971. Randomized models and the dilution and bioassay problems. Ann. Math. Stat. 42(1):141–6.
313. Frawley, W. H. 1974. Using the jackknife in testing dose responses in proportions near zero or one—revisited. Biometrics 30(3):539–46.
314. Freeman, P. R. 1970. Optimal Bayesian sequential estimation of the median effective dose. Biometrika 57:79–89.
315. Frome, E. L. and DuFrain, R. J. 1978. The analysis of dose-response curves for radiation-reduced chromosome aberrations. Paper presented at the Joint Stat. Conf., San Diego. Aug. 14–18, 1978.
316. Fry, F. E. J. 1947. Effects of the environment on animal activity. Ontario Fisheries Research Laboratory, No. 68. 1–62. (Reprinted in University of Toronto Studies, Biological Series No. 55 1947.)
317. Fryer, J. G. and Pethybridge, J. 1975. Maximum likelihood estimation for a special type of grouped data with an application to a dose-response problem. Biometrics 31(3):633–42.
318. Fryer, J. G. and Pethybridge, R. J. 1972. Maximum likelihood estimation of a linear regression function with grouped data. Applied Statistics 21:142–54.
319. Gaddum, J. H. 1933. Reports on Biological Standards. III. Methods of Biological Assay Depending on a Quantal Response. London: H. M. Stationery Office.
320. Gaddum, J. H. 1937. The quantitative effects of antagonistic drugs. J. of Physiology 89:547.
321. Gafarian, A. V. 1978. Confidence bands in multivariate polynomial regression. Technometrics 20:141–9.
322. Gaines, R. E. 1976. Analysis of radioimmunoassay data from international collaborative studies using weighted linear regression. Contributed paper, 9th International Biometric Conference, Boston.
323. Gallo, P. G. 1977. The age of menarche in some populations of the Veneto North Italy. Ann. Human Biology 4(2):179–81.
324. Gart, J. J. and Zweifel, J. R. 1967. On the bias of various estimators of the logit and its variance with application to quantal bioassay. Biometrika 54:181–7.
325. Garwood, F. 1941. The application of maximum likelihood to dosage-mortality curves. Biometrika 32:46–58.
326. Gaylor, D. W. 1980. The EDO1 study: Summary and conclusions. J. Environ. Pathol. Toxicol. 3:179–183.
327. Gaylor, D. W. and Kodell, R. L. 1980. Linear interpolation algorithm for low dose risk assessment of toxic substances. J. of Environmental Pathology and Toxicology 4:305–12.
328. Geary, R. C. 1930. The frequency distribution of the quotient of two normal variates. J. Roy. Stat. Soc. 93:442–6.
329. Geier, T. and Rohde, W. 1981. Comparison of four mathematical models for the calculation of radio immunoassay data of luteinizing hormone FSH and growth hormone. Endokrinologie 78(2–3):269–80.
330. Gill, J. A., Verts, B. J. and Christensen, A. G. 1970. Toxicities of DDT and some other analogs of DDT to pheasants. J. of Wildlife Management 34(1):223–6.
331. Gilman, L. and Rose, A. J. 1974. APL, An Interactive Approach, 2nd ed. New York: Wiley.
332. Goldenthal, E. I. 1971. A compilation of LD50 values in newborn and adults animals. Toxicol. Appl. Pharmacol. 18:185–207.
333. Goldstein, A. 1964. Biostatistics: An Introductory Text. New York: Macmillan.

334. Graca, J. G., Garst, E. L. and Lowry, W. E. 1957. Comparative toxicity of stable rare earth compounds. I. Effect of citrate complexing on stable rare earth chloride toxicity. American Medical Association, Archives of Industrial Health 15:9–14.
335. Grandage, A., Casady, R. B. and Lucas, H. L. 1954. The analysis of errors in assaying fecal androgens in the dairy cow. J. Dairy Sci. 37(1):72–80.
336. Grassivaro, G. P. and Franceshetti, M. M. 1980. Growth of children in Somalia. Human Biology 52(3):547–62.
337. Gray, H. L. and Schucany, W. R. 1972. The Generalized Jackknife Statistic. New York: Marcel Dekker.
338. Gray, H. L., Schucany, W. R. and Watkins, T. A. 1975. On the generalized jackknife and its relation to statistical differentials. Biometrika 63(3):637–42.
339. Grewal, R. S. 1952. A method for testing analgesics in mice. British J. of Pharm. and Chemotherapy 7:433–7.
340. Grey, L. D. 1973. A Course in APL/360 with Applications. Don Mills: Addison-Wesley.
341. Grizzle, J. E. and Allen, D. M. 1969. Analysis of growth and dose response curves. Biometrics 25:357–81.
342. Grotjan, H. E. and Steinberger, E. 1977. Radioimmunoassay and bioassay data processing using a logistic curve fitting routine adapted to a desk top computer. Computers in Biology and Medicine 7(2):159–63.
343. Guerrero, V. M. and Johnson, R. A. 1982. Use of the Box-Cox transformation with binary response models. Biometrika 69(2):309–14.
344. Gurland, J., Lee, I. and Dahm, P. A. 1960. Polychotomous quantal response in biological assay. Biometrics 16:382–98.
345. Halpern, M. 1970. An inverse estimation in linear regression. Technometrics 12:727–36.
346. Hamilton, M.A. 1979. Robust estimates of ED50. J. Amer. Stat. Assoc. 74:344–54.
347. Hamilton, M.A. 1980. Inference about ED50 using the trimmed Spearman-Kärber procedure—a Monto Carlo investigation. (Submitted to Commun. in Stat.)
348. Hamilton, M. A., Russo, R. C. and Thurston, R. V. 1977. Trimmed Spearman-Kärber method for estimating median lethal concentrations in toxicity bioassays. Environ. Sc. and Technol. 11(7):714–9. See corrections in 1978 issue, 12:417.
349. Harding, B. R., Thomson, R. and Curtis, A. R. 1973. A new mathematical model for fitting an HPL radioimmunoassay curve. J. of Clinical Pathology 26:973–6.
350. Harris, E. K. 1959. Confidence limits for the LD50 using the moving average-angle method. Biometrics 15:424–32.
351. Harris, R. J. 1975. A Primer of Multivariate Statistics. New York: Academic Press.
352. Harris, S. M. and Hubert, J. J. 1979. Ridit analysis and its application to epidemiological data. Tech. Rep. No. 1979–100, Dept. of Math. and Stat., Univ. of Guelph, 18p.
353. Hartley, H. O. 1975. Exact confidence regions in quantal bioassays. Biometrics 31(2):594.
354. Hartley, H. O. and Sielken, R. L. Jr. 1977. Estimation of "safe doses" in carcinogenic experiments. Biometrics 33(1):1–30.
355. Haseman, J. K. and Kupper, L. L. 1979. Analysis of dichotomous response data from certain toxicological experiments. Biometrics 35:281–93.
356. Hasselblad, V., Stead, A. G. and Creason, J. P. 1980. Multiple probit analysis with a nonzero background. Biometrics 36:659–63.
357. Hasselrot, T. B. 1975. Bioassay methods of the National Swedish Environment Protection Board. J. of Water Pollution Control 47(4):851–7.
358. Hazen, A. 1914. Storage to be provided in impounding reservoirs for municipal water supply. Trans. Amer. Soc. Civ. Engrs. 77:1539–669.
359. Healy, M. J. R. 1950. The planning of probit assays. Biometrics 6:424–31.
360. Healy, M. J. R. 1972. Statistical analysis of radioimmunoassay data. Biochemical J. 130:207–10.
361. Heltshe, J. F. 1983. The bootstrap, jackknife and rank transforms as applied to aquatic toxicity data. Technical report, presented at the Annual Toxicity Workshop, Halifax, N.S., Nov. 7–10.
362. Henderson, C. and Tarzwell, C. W. 1957. Bioassays for the control of industrial effluents. Sewage Industrial Wastes 29(9):1002–17.
363. Herbert, D. 1980. Bi-variate probit analysis of clinical dose-time data. Medical Physics 2(3):161.
364. Hersted, D. M. 1968. Determination of the relative nutritive value of protein. J. of Agriculture and Food Chemistry 16(2):190–5. (See reference to his 1965 paper in the J. of Nutrition.)
365. Hertzberg, J. O. 1975. On small sample properties of probit analysis. Proceedings of the 8th International Biometry Society, 153–64.
366. Hewlett, P. S. 1962. Toxicological studies on a beetle, *Alphitobius laevigatus* (F.). I. Dose-response relations for topically applied solutions of four toxicants in a non-volatile oil. Annals of Applied Biology 50:335–49.
367. Hewlett, P. S. 1963. Toxicological studies on a beetle, *Alphitobius laevigatus* (F.) III: The joint action of doses of each of four toxicants put on two parts of the body. Annals of Applied Biology 52:305–11.
368. Hewlett, P. S. 1969. Measurement of the potencies of drug mixtures. Biometrics 25:477–87.

369. Hewlett, P. S. and Plackett, R. L. 1950. Statistical aspects of the independent joint action of poisons, particularly insecticides. II. Examination of data for agreement with the hypothesis. Annals of Applied Biology 37:527–552.
370. Hewlett, P. S. and Plackett, R. L. 1956. The relation between quantal and graded responses to drugs. Biometrics 12:72–8.
371. Hewlett, P. S. and Plackett, R. L. 1959. A unified theory of quantal responses to mixtures of drugs: Noninteractive action. Biometrics 15:591–610.
372. Hewlett, P. S. and Plackett, R. L. 1961. Models for quantal responses to mixtures of two drugs. Quantitative Methods in Pharmacology, H. De Jonge, editor, p. 328–36. Amsterdam: North-Holland.
373. Hewlett, P. S. and Plackett, R. L. 1964. A unified theory of quantal responses to mixtures of drugs: Competitive action. Biometrics 20:566–75.
374. Hewlett, P. S. and Plackett, R. L. 1969. Models for quantal responses to mixtures of two drugs. Symposium Quarterly of Methods in Pharmacology.
375. Hewlett, P. S. and Plackett, R. L. 1979. The Interpretation of Quantal Responses in Biology. Baltimore: Univ. Park Press.
376. Hinkley, D. V. 1969. On the ratio of two correlated normal random variables. Biometrika 56:635–9.
377. Hinkley, D. V. 1970. Inference about the change-point in a sequence of random variables. Biometrika 57:1–17.
378. Hochberg, Y., Marom, I., Keret, R. and Peleg, S. 1983. On improved calibrations of unknowns in a system of quality-controlled assays. Biometrics 39(1):97–108.
379. Hodson, P. V. 1974. The effect of temperature on the toxicity of zinc to fish of the genus *Salmo*. Ph.D. thesis, Univ. of Guelph.
380. Hodson, P. V., Blunt, B. R. and Spry, D. J. 1978. pH-induced changes in blood level of lead-exposed rainbow trout (*Salmo gairdneri*). J. Fish. Res. Board Can. 35:437–45.
381. Hodson, P. V., Ross, C. W., Niimi, A. J. and Spry, D. J. 1976. Statistical considerations in planning aquatic bioassays. Report Environment Canada, Fisheries and Marine Service, Centre for Inland Waters, 28p.
382. Hodson, P. V. and Sprague, J. B. 1975. Temperature-induced changes in acute toxicity of zinc to Atlantic salmon (*Salmo salar*). J. Fish. Res. Board Can. 32:1–10.
383. Hoel, D. G. et al. 1975. Estimation of risks of irreversible, delayed toxicity. J. of Toxicology and Environmental Health 1:133–51.
384. Hoel, D. G. 1980. Incorporation of background responses dose-response models. Federation Proceedings 39:73–5.
385. Horn, H. J. 1956. Simplified LD50 (or ED50) calculations. Biometrics 12(3):311–22.
386. Horowitz, M. 1976. Application of bioassay techniques to herbicide investigations. Weed Research 16(4):209–15.
387. Hoskins, W. M. 1960. Use of the dosage-mortality curve in quantitative estimation of insecticide resistance. Miscellaneous Publications of the Entomology Society of America 2:85–91.
388. Hoskins, W. M. and Craig, R. 1962. Uses of bioassays in Entomology. Ann. Rev. Entomol. 7:437–64.
389. Howe, R. W. 1975. The estimation of developmental time by probit analysis. J. Stored Products Res. 11(2):121–2.
390. Hsi, B. P. 1969. The multiple sample up-and-down method in bioassay. J. Amer. Stat. Assoc. 64:147–62.
391. Hsia, J. C., Tam, J. C., Giles, H. G., Leung, C. C., Marcus, H., Marshman, J. A. and LeBlanc, A. E. 1976. Markers for detection of supplementation in narcotic programs—deuterium-labelled methadone. Science 93:498–500.
392. Huang, C. J. 1984. Estimation of seemingly unrelated tobit regressions. Paper for ASA Conference, Aug. 15, 1984, Philadelphia.
393. Huang, K. S. and Kurmeister, L. 1976. Optimal dosage under interactive drug combinations. Contributed paper, 9th International Biometric Conference, Boston.
394. Hubert, J. J. 1984. Multivariate growth curve modelling in bioassay. Paper given to the Dept. of Stat. and Act. Sc., Univ. of West. Ont., 18p.
395. Hubert, J. J. 1985. Assessment of pharmacological activity. To appear in Handbook of Biopharmaceutical Statistics in Human Drug Development. (K. E. Peace, ed.) New York: Marcel Dekker.
396. Hubert, J. J. and Carter, E. M. 1982. Confidence bands for time-dependent toxicity experiments. Paper for Joint Stat. Conf., Cincinnati, 23p.
397. Hubert, J. J. and Carter, J. J. 1985. PLASSAY: a BASIC program for microcomputers for parallel-line quantitative univariate bioassays. Tech. Rep. No. 1985–167, Dept. of Math. and Stat., Un. of Guelph, 18p.
398. Hubert, J. J. and Carter, E. M. 1985. SRAY: a BASIC program for microcomputers for slope-ratio assays. Tech. Rep. No. 1985–170. Dept. of Math. and Stat., Un. of Guelph, 19p.
399. Hubert, J. J., Fanning, L. P. and De France, C. 1984. Some software for an applied statistics course. Applied Math. Notes 9(1):20–31.
400. Hubert, J. J. and Hykle, D. J. 1984. A Study Guide to Biostatistics. Kendall/Hunt Publishing Company, Dubuque, Iowa, 179p.

401. Hubert, J. J. and Schock, J. P. 1984. PROBIT: an interactive BASIC program for probit analysis. Tech. Rep. No. 1984–165, Dept. of Math. and Stat., Univ. of Guelph, p14.
402. Hung, W. H. 1985. Microcomputer Application in Aquatic Toxicity Testing, Ph.D. Thesis, Dept. of Environ. Health Sciences, Un. of Michigan
403. Hui, S. L. and Rosenberg, S. N. 1985. Multivariate slope ratio assay with repeated measurements. Biometrics 41(1):11–18.
404. Hunter, W. G. and Lamboy, W. F. 1981. A Bayesian analysis of the linear calibration problem (with discussion). Technometrics 23(4):323–50.
405. Irwin, J. O. 1937. Statistical methods applied to biological assays. J. Roy. Stat. Soc. Supp. 4(1):1–48.
406. Irwin, J. O. 1950. Biological assays with special reference to biological standards. J. of Hygiene, Cambridge 48:215–38.
407. Irwin, J. O. and Cheeseman, E. A. 1939. On the maximum likelihood method of determining dosage-response curves and approximations to the median effective dose, in cases of a quantal response. J. Roy. Stat. Soc. Supp. 6(2):174–85.
408. Irwin, J. O. and Cheesman, E. A. 1939. On an approximate method of determining the median effective dose and its error in the case of a quantal response. J. of Hygiene 39:574–80.
409. Jaeger, R. J. 1976. Kepone chronology. (Letter) Science 193(4248):94.
410. James, B. R. and James, K. L. 1983. On the influence curve for quantal bioassay. J. Statistical Planning and Inference (3):331–45.
411. Janardan, K. G. and Schaeffer, D. J. 1980. A probability model for the distribution of compounds in the aquatic environment. Tech. Rep. No. 81–14, Dept. of Stat., Univ. of Pittsburgh.
412. Jenicek, M. and Demirjian, A. 1974. Age at menarche in French Canadian urban girls. Ann. Human Biology 1(3):339–46.
413. Jensen, A. L. 1972. Standard error of LC50 and sample size in fish bioassays. Water Research 60(1):85–9.
414. Jerne, N. R. and Wood, E. C. 1949. The validity and meaning of the results of biological assays. Biometrics 5:275–99.
415. Johnson, E. A. and Brown, B. W. Jr. 1961. The Spearman estimator for serial dilution assay. Biometrics 17:79–88.
416. Jordan, G. W. 1971. Basis for the probit analysis of an interferon placque reduction assay. J. of General Virology 4(1):49–61.
417. Jørgensen, B. 1983. Maximum likelihood estimation and large-sample inference for generalized linear and nonlinear regression models. Biometrika 70(1):19–28.
418. Kalbfleisch, J. D., Krewski, D. R. and Van Ryzin, J. 1983. Dose-response models for time-to-response toxicity data. Canadian J. of Statistics. 11(1):25–49.
419. Kalbfleisch, J. D. and Prentice, R. L. 1980. The Statistical Analysis of Failure Time Data. New York: Wiley.
420. Kärber, G. 1931. Beitrag zur kollektiven Behandlung pharmakologischer Reihenversuche. Arch. Exp. Path. Pharmak. 162:480–7.
421. Kemp, P. L. 1979. Nonlinear estimation of radioimmunoassay data. Contributed paper, Joint Stat. Meetings, New Orleans, 17p.
422. Kempthorne, O. 1952. The Design and Analysis of Experiments. New York: Wiley.
423. Kershaw, C. D. 1983. Sequential estimation for binary response. Ph.D. Thesis, Univ. of Edinburgh.
424. Kershaw, C. D. 1985. Asymptotic properties of $\bar{w}$, an estimator of the ED50 suggested for use in up-and-down experiments in bio-assay. Annals of Statistics 13(1):85–94.
425. Khatri, C. G. 1966. A note on a MANOVA model applied to problems in growth curves. Ann. Inst. Statist. Math. 18:75–86.
426. Khatri, C. G. 1973. Testing some covariance structures under a growth curve model. J. of Multivariate Analysis 3:102–16.
427. Khatri, C. G. and Shah, K. R. 1979. Estimation in non-linear growth curves. Tech. Rep. No. 79–04, Faculty of Math., Univ. of Waterloo.
428. Kido, R. C., Asano, C. and Yamamoto, K. 1956. The bioassay of potency of short acting narcotics. Annual Report of the Shionogi Research Laboratory, (6), 203–11. (In Japanese)
429. Kiefer, N. M. 1982. Testing for dependence in multivariate probit models. Biometrika 69(1):161–6.
430. Klein, J. P. 1981. A life table approach to the dose-response problem. Tech. Rep. No. 1981–226, Dept. of Stat., Ohio State Univ.
431. Knudsen, L. F. and Curtis, J. M. 1947. The use of the angular transformation in biological assays. J. Amer. Stat. Assoc. 42:889–902.
432. Kolakowski, D. and Bock, R. D. 1981. A multivariate generalization of probit analysis. Biometrics 37:541–51.
433. Könemann, W. H. 1979. Quantitative structure-activity relationships for kenetics and toxicity of aquatic pollutants and their mixtures in fish. Ph.D. Thesis, Univ. of Utrecht.
434. Konishi, S. 1978. Asymptotic expansions for the distributions of statistics based on a correlation matrix. Can. J. Stat. 6:49–56.
435. Kooijman, S. A. L. M. 1981. Parametric analyses of mortality rates in bioassays. Water Research 15(1):107–20.

436. Kooijman, S. A. L. M. 1983. Statistical aspects of the determination of mortality rates in bioassays. Water Research 17(7):749–59.
437. Kooijman, S. A. L. M., Hanstveit, A. O. and Oldersma, H. 1983. Parametric analyses of population growth in bio-assays. Water Research 17(5):527–38.
438. Koopman, P. A. R. 1984. Confidence intervals for the ratio of two binomial proportions. Biometrics 40(2):513–8.
439. Korn, E. L. 1982. Confidence bands for isotonic dose-response curves. Applied Statistics 31(1):59–63.
440. Koshiver, J. and Moore, D. 1979. LOGIT: a program for dose response analysis. Comput. Programs Biomed. 10(1):61–5.
441. Kraft, C. H., Olkin, I. and Van Eeden, C. 1972. Estimation and testing for differences in magnitude or displacement in the mean vectors of two multivariate normal populations. Annals of Mathematical Statistics 43(2):455–67.
442. Kraft, C. H. and Van Eeden, C. 1964. Bayesian bioassay. Annals of Mathematical Statistics 35:886–90.
443. Krewski, D., Smythe, R. and Colin, D. 1985. A two-stage procedure for incorporating historical controls information in testing for trend in quantal response toxicity data. Symposia in Statistics Proceedings, Univ. of Western Ontario.
444. Krutchkoff, R. G. 1967. Classical and inverse regression methods of calibration. Technometrics 9:425–39.
445. Krutchkoff, R. G. 1969. Classical and inverse regression methods of calibration in extrapolation. Technometrics 11:605–8.
446. Kudo, A. and Furukawa, N. 1958. A model in probit analysis. Bulletin of Mathematical Statistics 8:1–7.
447. Kulshrestha, A. C. 1969. Modified incomplete block bioassay designs (abstract). Proc. Indian Sc. Cong., 56th Session, 3:33.
448. Kulshrestha, A. C. 1971. Incomplete block designs and their applications to bioassays. Ph.D. Thesis, Inst. of Advanced Studies, Meerut Un., India.
449. Kumaraguru, A. K. and Beamish, F. W. H. 1981. Lethal toxicity of permethrin (NRDC-143) to rainbow trout, *Salmo gairdneri*, in relation to body weight and water temperature. Water Research 15:503–5.
450. Kuo, L. 1980. Bayesian bioassay design. Tech. Rep. No. 101, Dept. of Stat., Univ. of Michigan.
451. Kuo, L. 1983. Bayesian bioassay design. Annals of Stat. 11(3):886–95.
452. Kupper, L. L. and Haseman, J. K. 1978. The use of a correlated binomial model for the analysis of certain toxicological experiments. Biometrics 34:69–76.
453. Lampkin, H. and Ogawa, J. 1976. Estimation of distribution parameters in time mortality trials. An example of time mortality analysis. Can. J. Stat. 4(1):65–93.
454. Larsen, R. I., Gardner, D. E. and Coffin, D. L. 1979. An air quality data analysis system for interrelating effects, standards, and needed source reductions: Part 5, NO2 mortality in mice. J. Air Pollution Control Assoc., 39:133–7.
455. Laska, E., Kushner, H. B. and Meisner, M. 1985. Multivariate bioassay (Reader Response). Biometrics 41(2):231–8.
456. Lawley, D. N. 1943. On problems connected with item selection and test construction. Proc. Roy. Soc., Edinburgh, 61:273–87.
457. Lawrence, C. J. 1974. Sequential estimation in bioassay. Biometrics 30(1):216.
458. Lee, J. Y. and Rohde, C. A. 1976. Likelihood techniques for some discrete distributions. Tech. Rep. No. 504, Dept. of Biostatistics, Johns Hopkins Univ.
459. Lee, M. L. and Chang, P. C. 1981. Some nearest neighbor estimators for the inverse regression problem. UCLA Tech. Rep., 92p.
460. Lee, M. L. and Chang, P. C. 1984. A nearest-neighbor estimation scheme for radioimmunoassays. Paper for Intern. Biometric Conf., Aug. 6.
461. Lee, Y. K. 1974. A note on Rao's reduction of Potthoff and Roy's generalized linear model. Biometrika 61:349–51.
462. Lejarraga, H., Sanchirico, F. and Cusminsky, M. 1980. Age of menarche in urban Argentinian girls. Ann. Human Biology 7(6):579–82.
463. Lesnick, T. G. 1986. Joint Action Toxicity Models. M.Sc. Thesis, Department of Mathematics and Statistics, Un. of Guelph.
464. Lett, P. F. and Beamish, F. W. M. 1975. System simulation of the predatory activities of sea lampreys (*Petromyzon marinus*) on lake trout (*Salvelinus namaycush*). J. Fish. Res. Board Can. 32(5):623–31.
465. Lett, P. F., Farmer, G. J. and Beamish, F. W. 1976. Effect of copper on some aspects of the bioenergetics of rainbow trout (*Salmo gairdneri*). J. Fish. Res. Board Can. 33(6):1335–42.
466. Litchfield, J. T. Jr. 1949. A method for rapid graphic solution of time-percent effect curves. J. Pharmac. and Exp. Ther. 97:399–408.
467. Litchfield, J. T. Jr. and Fertig, J. W. 1941. On a graphic solution of the dosage-effect curve. Johns Hopkins Hosp. Bull. 69:276–86.
468. Litchfield, J. T. Jr. and Wilcoxon, F. 1949. A simplified method of evaluating dose-effect experiments. J. Pharmac. and Exp. Ther. 96:99–113.
469. Litchfield, J. T. Jr. and Wilcoxon, F. 1953. The reliability of graphic estimates of relative potency from dose-percent effect curves. J. Pharmac. and Exp. Ther. 108:18–25.

470. Little, R. E. 1968. A note on estimation for quantal response data. Biometrika 55(3):578–9.
471. Little, R. E. 1974. A mean square error comparison of a certain median response estimates for the up-and-down method with small samples. J. Amer. Stat. Assoc. 69:202–6.
472. Little, R. E. 1974. The up-and-down method for small samples with extreme value response distributions. J. Amer. Stat. Assoc. 69(347):803–6.
473. Little, R. E. 1975. The up-and-down method for small samples with two specimens "in series". J. Amer. Stat. Assoc. 70:846–51.
474. Lloyd, R. 1960. The toxicity of zinc sulphate to rainbow trout. Annals of Applied Biology 48(1):84–94.
475. Lukasewycz, O. A., Martinez, D. and Murphy, W. H. 1975. Immune mechanisms in leukemia: evaluation of immunocompetent cell populations. J. of Immunology 114(5):1491–6.
476. Lu, F. C. 1985. Basic Toxicology. McGraw-Hill: Toronto.
477. Lund, J. W. G. and Jaworski, G. H. M. 1971. A technique for bioassay of fresh water with special reference to algae ecology. Acta Hydrobiologica 13(3):235–49.
478. Machado, V. S., Nonnecke, I. L. and Phatak, S. C. 1978. Bioassay to screen tomato seedlings for tolerance to metribuzin. Can. J. of Plant Science 58(3):823–8.
479. Maciorowski, A. F., Little, L. W., Raynor, L. F., Sims, R. C. and Sims, J. L. 1983. Bioassays—procedures and results. J. of Water Pollution Control 55:801–16.
480. Magnus, A., Mielke, P. W. and Copenhaver, T. W. 1977. Closed expressions for the sum of an infinite series with application to quantal response assays. Biometrics 33(1):221–4.
481. Malcolm, S. A. 1978. Analysis of grouped survival times and threshold estimation in acute toxicity studies. Master's Thesis, Dept. of Math. and Stat., Univ. of Guelph.
482. Malcolm, S. A., Pursey, S. D. and Hubert, J. J. 1976. Preliminary report on computer functions for bioassay. Statistical Series No. 1976–41. Dept. of Math. and Stat., Univ. of Guelph, 58p.
483. Malik, H. J. and Mullen, K. 1973. A First Course in Probability and Statistics. Don Mills: Addison-Wesley.
484. Malley, J. D. 1982. Simultaneous confidence intervals for ratios of normal means. J. Amer. Stat. Assoc. 77(377):170–6.
485. Mancini, J. L. 1983. A method for calculating effects, on aquatic organisms, of time varying concentrations. Water Research 17(10):1355–62.
486. Mantel, N. 1973. Quantal bioassay with one animal at a dose level. Biometrics 29(1):225–6.
487. Mantel, N. and Bryan, W. R. 1961. "Safety" testing of carcinogenic agents. J. Nat. Cancer Inst. 27:455–70.
488. Mantel, N. and Greenhouse, S. W. 1967. Equivalence of maximum likelihood and the method of moments in probit analysis. Biometrics 23:154–7.
489. Mantel, N. and Hilgar, A. G. 1961. Effects of non linearity on subjective and objective analysis of hormone assay data. Cancer Chemotherapy Reports 11:61–8.
490. Mantel, N. and Schneiderman, M. A. 1975. Non-parametric interval estimation of relative potency for dilution assays, including the case of non-monotone dosage response curves. Biometrics 31(3):619–32.
491. Margolin, B. 1981. Experiences in statistically validating new bioassays. Contributed paper for Biometric Society, Detroit.
492. Margolin, B. N., Kaplan, N. and Zeiger, E. 1981. Statistical analysis of the Ames Salmonella-microsome test. Proc. Nat. Acad. Sciences, 78(6):3779–83.
493. Martin, J. T. 1942. The problem of the evaluation of rotenone-containing plants. VI. The toxicity of 1-elliptone and of poisons applied jointly, with further observations on the rotenone equivalent method of assessing the toxicity of derris root. Annals of Applied Biology 29:69–81.
494. Matterson, L. D. and Pudelkiewicz, W. J. 1974. Relative potency of several forms of alpha-Tocopherols in the chick liver storage bioassay. J. of Nutrition 104:79–83.
495. McArthur, J. W., Ulfelder, H. and Finney, D. J. 1966. A flexible computer program for the composite analysis of symmetrical biologic assays of parallel-line type. J. Pharmac. and Exp. Ther. 153:573–80.
496. McCarty, L. S., Henry, J. A. C. and Houston, A. H. 1978. Toxicity of cadmeum to goldfish, *Carassius auratus,* in hard and soft water. J. Fish. Res. Board Can. 35:35–42.
497. McLeese, D. W. 1974. Toxicity of phosphamidon to American lobsters (*Homarus americanus*) held at 4 and 12 C. J. Fish. Res. Board Can. 31(9):156–8.
498. McLeese, D. W. 1974. Toxicity of copper at two temperatures and three salinities to the American lobster (*Homarus americanus*). J. Fish. Res. Board Can. 31(12):1949–52.
499. McLeish, D. and Tosh, D. 1983. The estimation of extreme quantiles in logit bioassay. Biometrika 70(3):625–32.
500. Meinert, C. L. and McHugh, R. B. 1968. The biometry of an isotope placement immunologic microassay. Mathematical Sciences 2:319–38.
501. Merrill, A. S. 1928. The frequency distribution of an index when both the components follow the normal law. Biometrika 20:53–63.
502. Meynell, G. G. 1957. Inherently low precision of infectivity titrations using a quantal response. Biometrics 13:149–63.
503. Mielke, P. W. and Flueck, J. A. 1976. Distributions and moments of ratios for selected bivariate probability functions. Abstracts, Inst. Math. Statist. Bull.

504. Milcer, C. 1968. Age at menarche of girls in Wroclaw, Poland, in 1966. Human Biology 40:249–59.
505. Milicer, H. and Szczotka, F. 1966. Age at menarche in Warsaw girls in 1965. Human Biology 38:199–203.
506. Miller, L. C., Bliss, C. I. and Braun, H. A. 1939. The assay of digitalis. I. Criteria for evaluating various methods using frogs. J. Amer. Pharm. Assoc. 28:644–57.
507. Miller, L. C. and Tainter, M. L. 1944. Estimation of the ED50 and its error by means of logarithmic-probit graph paper. Proc. Soc. of Experimental Biology and Medicine 57:261–4.
508. Miller, R. G. 1964. A trustworthy jackknife. Annals of Mathematical Statistics 35:1594–605.
509. Miller, R. G. 1973. Nonparametic estimators of the mean tolerance in bioassay. Biometrika 60(3):535–42.
510. Miller, R. G. 1974. The jackknife—a review. Biometrika 61(1):1–15.
511. Miller, R. G. 1974. An unbalanced jackknife. Annals of Statistics 2(5):880–91.
512. Miller, R.G., and Halpern, J. W. 1980. Robust estimators for quantal bioassay. Biometrika 67(1):103–10.
513. Minder, C. E. and Whitney, J. B. 1975. A likelihood analysis of the linear calibration problem. Technometrics 17(4):463–71.
514. Minkin, S. 1982. Assessing the quadratic approximation to the log-likelihood function in non-normal linear models. Tech. Rep. Dept. of Math. and Stat., Univ. of Guelph.
515. Moore, R. H. and Zeigler, R. K. 1967. The use of non-linear regression methods for analyzing sensitivity and quantal response data. Biometrics 23:563–6.
516. Moran, P. A. P. 1954. The dilution assay of viruses. I. J. of Hygiene, Cambridge 52:189–93.
517. Moran, P. A. P. 1954. The dilution assay of viruses. II. J. of Hygiene, Cambridge 52:444–6.
518. Morgan, B. J. T. 1983. Observations on quantit analysis. Biometrics 39(4):879–86.
519. Morton, N. E. and Yee, S. 1971. Bioassay of kinship. Theor. Pop. Biol. 2(4):507–24.
520. Mosteller, F. 1946. On some useful 'inefficient' statistics. Annals of Mathematical Statistics 17:377–408.
521. Muthén, B. 1979. A structural probit analysis model with latent variables. J. Amer. Stat. Assoc., 74:807–11.
522. Myers, L. E., Sexton, N. H., Southerland, L. I. and Wolff, T. J. 1981. Regression analysis of Ames test data. Envir. Mutagenesis 3:575–86.
523. Nagasawa, S. 1959. Biological assay of insecticidal residues. Ann. Rev. Entomol. 4:319–42.
524. Narasimhan, C. 1984. Tobit analysis of coupon usage. Paper for the ASA Conf., Aug., Phil.
525. Narayana, T. V. 1953. Sequential procedures in probit analysis. Ph.D. Thesis, Univ. of North Carolina.
526. Naylor, A. F. 1964. Comparisons of regression constants fitted by maximum likelihood to four common transformations of binomial data. Annals of Human Genetics 27:241–6.
527. Nelder, J. A. 1968. Weighted regression, quantal response data, and inverse polynomials. Biometrics 24:979–85.
528. Nelder, J. A. and Mead, R. 1965. A simplex method for function minimization. Computing Journal 7:308–13.
529. Niimi, A. J. and Lattam, Q. N. 1976. Relative toxicity of organic compounds of selenium to newly hatched zebrafish (*Brachydanio rerio*). Can. J. Zoology 54:501–9.
530. Ochi, Y. and Prentice, R. L. 1984. Likelihood inference in a correlated probit model. Biometrika 71(3):531–44.
531. Ochi, Y. and Prentice, R. L. 1984. The correlated probit regression for count data with extra-binomial variation. Paper for Int. Biometrics Conference, Aug. 4, 1984.
532. Odell, P. L. 1961. An empirical study of three stochastic approximation techniques applicable to sensitivity testing. NAVWEPS Report No. 7837, 35p.
533. Olkin. I. 1985. Multivariate normal models with multiplicative factors. To appear in Symposia in Statistics, Univ. of Western Ont.
534. O'Neil, J. D., Brown, D. J. and Forney, R. B. 1979. A new approach to an old method: a basic program for the calculation of the ED50 or LD50 using probit analysis. Toxicology and Applied Pharmacology 48(1–2):102.
535. Onon, E. O. 1979. Purification and partial characterization of the exotoxin of *Corynebacterium ovis*. Biochemistry J. 177:181–6.
536. Ozburn, G. W. and Morrison, F. O. 1964. The selection of DDT-tolerant strain of mice and some characteristics of that strain. Can. J. Zoology 42:519–26.
537. Parker, R. E. 1973. Introductory Statistics for Biology. Toronto: Macmillan.
538. Patwary, K. M. 1960. Error and non-error models in bio-assay. Ph.D. Thesis. American Univ., Washington, D.C. (See pages 11–40.)
539. Patwary, K. M. and Haley, K. D. C. 1967. Analysis of quantal response assays with dosage errors. Biometrics 23:747–60.
540. Pekary, A. E. 1980. Parallel line and relative potency analysis of bioassay and radioimmunoassay data using a desk top computer. Computers in Biology and Medicine 9(4):355–62.
541. Pelczar, M. J. and Reid, R. D. 1972. Microbiology, 3rd ed. Toronto: McGraw-Hill.
542. Pendergast, J. F. and Broffitt, J. D. 1982. Robust estimation in growth curve models. Tech. Rep. No. 104, 22p. Dept. of Stat., Univ. of Florida.
543. Peto, R. and Lee, P. 1973. Weibull distributions for continuous carcinogenesis experiments. Biometrics 29:457–70.

544. Peto, R., Lee, P. M. and Paige, W. S. 1972. Statistical analysis of the bioassay of continuous carcinogens. British J. of Cancer 26:258–61.
545. Peto, S. 1953. A dose-response equation for the invasion of microorganisms. Biometrics 9:320–35.
546. Petrusz, P., Diczfalusy, E. and Finney, D. J. 1971. Bioimmunoassay of gonadotrophins. 1. Theoretical considerations. Acta Endocrinologica 67:40–6.
547. Petrusz, P., Diczfalusy, E. and Finney, D. J. 1971. Bioimmunoassay of gonadotrophins. 2. Practical aspects and tests of additivity. Acta Endocrinologica 67:46–62.
548. Pfaeffle, W. O. 1958. Biological assays of Guthion residues on alfalfa treated for insect control. Master's thesis, Iowa State College, Ames, Iowa.
549. Pilli, A. 1983. AQUIRE: Aquatic toxicity information retrieval. Tech. Rep. for the Annual Toxicity Workshop, Halifax, Nov. 7–9.
550. Pizzi, M. 1950. Sampling variation of the fifty percent end-point, determined by the Reed-Muench (Behrens) method. Human Biology 22:151–90.
551. Plackett, R. L. and Hewlett, P. S. 1952. Quantal responses to mixtures of poisons (with discussion). J. Roy. Stat. Soc., B, 14:141–63.
552. Plackett, R. L. and Hewlett, P. S. 1963. A unified theory for quantal responses to mixtures of drugs: the fitting to data of certain modes for two non-interactive drugs with complete positive correlation of tolerances. Biometrics 19:517–31.
553. Plackett, R. L. and Hewlett, P. S. 1967. A comparison of two approaches to the construction of models for quantal responses to mixtures of drugs. Biometrics 23:27–44.
554. Portier, C. and Hoel, D. 1983. Low-dose rate extrapolation using the multistage model. Biometrics 39(4):897–906.
555. Potthoff, R. F. and Roy, S. N. 1964. A generalized multivariate analysis of variance model useful especially for growth curve problems. Biometrika 51:313–26.
556. Prairie, R. R. 1967. Probit analysis as a technique for estimating the reliability of a simple system. Technometrics 9:197–203.
557. Prentice, R. L. 1976. A generalization of the probit and logit methods for dose response curves. Biometrics 32:761–8.
558. Preisler, H. K. 1985. Random effects models for combining results in Entomological bioassay studies. I.M.S. Conference (San Luis, Ca) Abstract in the Bulletin 14(4):172. (USFS, PSW, T.R., 14p.)
559. Preston, E. J. 1952. A graphical method for analysis of statistical distributions into normal components. Biometrika 40:460–4.
560. Pugsley, L. J. 1946. The application of the principles of statistical analysis to the biological assay of hormones. Endocrinology 39:1–76.
561. Pursey, S. D. 1977. Non-parametric procedures in quantal bioassays. M.Sc. Thesis, Dept. of Math. and Stat., Univ. of Guelph.
562. Pursey, S. D. and Hubert, J. J. 1977. SRAY: An APL program for slope ratio quantitative bioassays. Tech. Rep. No. 1977–65, Dept. of Math. and Stat., Univ. of Guelph, 19p.
563. Raab, G. M. 1981. Estimation of a variance function, with application to immunoassay. Applied Statistics 30(1):32–40.
564. Rai, K. and Van Ryzin, J. 1985. A dose-response model for teratological experiments involving quantal responses. Biometrics 41(1):1–10.
565. Raktoe, B. L. and Hubert, J. J. 1979. Basic Applied Statistics. New York: Marcel Dekker.
566. Ramsey, F. L. 1972. A Bayesian approach to bioassay. Biometrics 28(3):841–58.
567. Rao, C. R. 1954. Estimation of relative potency from multiple response data. Biometrics 10:208–20.
568. Rao, C. R. 1959. Some problems involving linear hypothesis in multivariate analysis. Biometrika 46:49–58.
569. Rao, C. R. 1965. The theory of least squares when the parameters are stochastic and its application to the analysis of growth curves. Biometrika 52:447–58.
570. Rao, P. V. and Littell, R. C. 1971. Robust estimation of shift parameters on Kolmogorov-Smirnov statistics. Tech. Rep. No. 30, Dept. of Stat., Univ. of Florida, 19p.
571. Rao, P. V. and Littell, R. C. 1972. Confidence intervals for shift parameters based on Kolmogorov-Smirnov statistics. Tech. Rep. No. 30A, Dept. of Stat., Univ. of Florida, 25p.
572. Rao, P. V. and Littell, R. C. 1976. An estimator of relative potency. Commun. in Stat. A 5(2):183–9.
573. Rao, P. V., Schuster, E. F. and Littell, R. C. 1975. Estimation of shift and center of symmetry based on Kolmogorov-Smirnov statistics. Annals of Statistics 3(4):862–73.
574. Reed, L. J. and Muench, H. 1938. A simple method of estimating fifty percent endpoints. Amer. J. Hygiene 27:493–7.
575. Reid, N. 1981. Estimating the median survival time. Biometrika 68:601–8.
576. Reid, R. M. 1974. Bioassay of inbreeding in India. Amer. J. Physical Anthropology 41(3):500.
577. Reinsel, G. 1982. Multivariate repeated-measurement or growth curve models with multivariate random-effects covariance structure. J. Amer. Stat. Soc. 77(377):190–6.
578. Rich, R. R., Lang, K. T. and Jones, A. L. 1974. The interpretation of bioassay laboratory results. Health Physics 27(6):630.

579. Robbins, H. and Monro, S. 1951. A stochastic approximation method. Annals of Mathematical Statistics 22:400–7.

580. Robel, E. J. 1978. Use of the slope ratio bioassay for estimating amino-acids in soybean meal for chicks. Poultry Science 57(4):1183.

581. Roberts, D. F., Chinn, S., Girija, B. and Singh, H. D. 1977. A study of menarcheal age in India. Ann. Human Biology 4(2):171–7.

582. Roberts, M. H. 1977. Bioassay procedures for marine phyto plankton with special reference to chlorine. Chesapeake Science 18(1):137–9.

583. Robertson, J. L., Russell, R. M. and Savin, N. E. 1981. POLO2: A computer program for multiple probit or logit analysis. Bull. Ent. Soc. Amer. 27(3):210–1.

584. Robertson, J. L., Savin, N. E. and Russell, R. M. 1981. Weight as a variable in the response of the western spruce budworm to insecticides. J. of Economic Entomology (in press).

585. Robertson, P. L. and Orton, C. J. 1971. Techniques for pheromone bioassay studies of ants. Bulletin of Entomological Research 61(2):283–91.

586. Robinson, J.A. 1976. Choosing an optimal dose: a sequential prediction interval approach. Contributed paper, Biometric Conference, Boston.

587. Robson, D. S., Hildreth, B. P., Atkinson, G. F., Carmicheal, L. E., Barnes, F. D., Pakkala, B. and Baker, J. A. 1961. Standardization of quantitative serological tests. 65th Annual Proceedings of the U.S. Livestock Sanitary Association, Oct, 74–78.

588. Robyn, C. and Diczfalusy, E. 1968. Bioassay of antigonadatrophic sera. 2. Assay of the human chorionic gonadotrophin (HCG) and luteinizing hormone (LH) neutralising potencies. Acta Endocrinologica 59:261–76.

589. Robyn, C., Diczfalusy, E. and Finney, D. J. 1968. Bioassay of anti-gonadatrophic sera. 1. Statistical considerations and general principles. Acta Endocrinologica 58:593–9.

590. Rodbard, D. 1971. Statistical aspects of radioimmunoassays. In: Principles of Competitive Protein Binding Assays. p. 204–59. Philadelphia: Lippincott. (W. D. Odell, ed.)

591. Rodbard, D. 1978. Data processing for radioimmunoassay: an overview. Current Topics in Clinical Chemistry 3:477–94.

592. Rodbard, D., Bridson, W. and Rayford, P. L. 1969. Rapid calculation of radioimmunoassay results. J. of Lab. Clinical Medicine 74:770–81.

593. Rodbard, D. and Cooper, T. A. 1970. A model for prediction of confidence limits in radioimmunoassays and competitive protein assays. Proc., Symposium on Radioisotopes in Medicine, Vienna, 659–74.

594. Rodbard, D. and Frazier, G. R. 1974. Statistical analysis of radioligand assay data. In: In Vitro Procedures with Radioisotopes in Medicine.

595. Rodbard, D. and Hutt, D. M. 1974. Statistical analysis of radioimmunoassays and immunoradiometric (labelled antibody) assays: a generalized weighted, iterative, least squares method for logistic curve fitting. In: Radioimmunoassay and Related Procedures in Clinical Med. and Res.

596. Rodbard, D., Lennox, H. L., Wray, H. L. and Ramseth, D. 1976. Statistical characterization of the random errors in the radioimmunoassay dose-response variable. Clinical Chemistry 22:350–8.

597. Rodbard, D. and Lewald, J. E. 1970. Computer analysis of radioligand assay and radioimmunoassay data. Acta Endoc. Supp. 147:79–103.

598. Rosenblatt, L. and Rosenblatt, J. 1973. Simplified Fortran Programming: with Companion Problems. Don Mills: Addison-Wesley.

599. Rousel, J. S., Saba, F. S., Reynolds, T. H., Forgash, A. J., Brazzel, J. R., Burkhardt, C. C. and Collins, W. J. 1972. Standard methods for detection of insecticide resistance in *Diabrotica* and *Hypera* beetles. Bull. Ent. Soc. Amer. 18:179–82.

600. Roy, S. N. and Potthoff, R. F. 1958. Confidence bounds on vector analogues of the 'ratio of means' and the 'ratio of variances' of two correlated normal variates and some associated tests. Annals of Mathematical Statistics 29:829–41.

601. Russell, R. M., Robertson, J. L. and Savin, N. E. 1977. POLO: a new computer program for probit analysis. Bull. Ent. Soc. Amer. 23:209–13.

602. Russell, R. M. and Robertson, J. L. 1979. Programming probit analysis. Bull. Ent. Soc. Amer. 25(3):191–2.

603. Russo, R. C., Smith, C. E. and Thurston, R. V. 1974. Acute toxicity of nitrate to rainbow trout (*Salmo gairdneri*). J. Fish. Res. Board Can. 31(10):1653–5.

604. Saito, H., Ishida, S., Yasuda, J., Kurokawa, M., Nagai, T. and Takahashi, T. 1979. Application of the parallel line bioassay method to quantitative determination of hepatitis B surface antigenin radio immunoassay. Japanese J. of Medical Science and Biology 32(1):47–52.

605. Salsburg, D. 1971. Testing dose responses on proportions near zero or one with the jackknife. Biometrics 27(4):1035–41.

606. Salsburg, D. S. 1975. Estimating differences in variance when comparing two methods of assay. Technometrics 17(3):381–2.

607. Sampford, M. R. 1952. The estimation of response time distributions. I. Fundamental concepts and general methods. Biometrics 8:13–32.

608. Sanders, H. O. 1970. Pesticide toxicities to tadpoles of the Western Chorus frog (*Psedacris triseriata*) and Fowler's toad (*Bufo woodhousii Fowleri*). Copeia 2:246–51.
609. Savin, N. E., Robertson, J. L. and Russell, R. M. 1977. A critical evaluation of bioassay in insecticide research: likelihood ratio tests of dose-mortality regression. Bull. Ent. Soc. Amer. 23:257–66.
610. Savin, N. E., Robertson, J. L. and Russell, R. M. 1981. The effect of insect weight on lethal dose estimates. To appear, Bull. Ent. Soc. Amer.
611. Saw, J. G. 1970. Letter to the editor. Technometrics 12:937.
612. Schaeffer, D. J., Glave, W. R. and Janardan, K. G. 1982. Multivariate statistical methods in toxicology. III. Specifying joint toxic interaction using multiple regression analysis. J. Tox. Envir. Health 9:705–18.
613. Schang, M. J. and Hamilton, R. M. G. 1981. Comparison of two direct bioassays using adult cockerels for estimating the available energy content of 13 feeding stuffs. Poultry Science 60(7):1726–7.
614. Schesser, J. H. and Bulla, L. A. Jr. 1978. Toxicity of *Bacillus thuringiensis* spores to the tobacco hornworm, *Manduca sexta*. Appl. Environ. Micro. 35(1):121–3.
615. Schesser, J. H., Kramer, K. J. and Bulla, L. A. Jr. 1977. Bioassay for homogeneous parasporal crystal of *Bacillus thuringiensis* using the tobacco hornworm, *Maduca sexta*. Appl. Environ. Micro. 33:870–80.
616. Schmetterer, L. 1960. Stochastic approximation. Proc. of the 4th Berkeley Symposium on Math. Stat. and Prob., 587–609.
617. Schmoyer, R. L. 1984. Sigmoidally constrained maximum likelihood estimation in quantal bioassay. J. Amer. Stat. Assoc. 79:448–53.
618. Schneiderman, M. A. 1974. Safe dose? Problem of the statistician in the world of trans-science. J. of Wash. Academy of Sciences 64:68–78.
619. Schoofs, G. M. and Willhite, C. C. 1984. A probit analysis program for the personal computer. J. Applied Toxicology 4(3):141–4.
620. Segreti. A. C. and Munson, A. E. 1981. Estimation of the median lethal dose when responses within a litter are correlated. Biometrics 37(1):153–6.
621. Sen, P. K. 1963. On the estimation of relative potency in dilution (direct) assays by distribution-free methods. Biometrics 19:532–52.
622. Sen, P. K. 1964. Tests for the validity of the fundamental assumption in dilution (direct) assays. Biometrics 20:770–84.
623. Sen, P. K. 1965. Some further applications of non-parametric methods in dilution (direct) assays. Biometrics 21:799–810.
624. Seth, A. and Hubert, J. J. 1977. PLASSAY: an APL program for parallel line quantitative bioassays. Tech. Rep. No. 1977–60. Dept. of Math. and Stat., Univ. of Guelph, 26p.
625. Sharot, T. 1976. Sharpening the jackknife. Biometrika 63(20):315–21.
626. Sharot, T. 1976. The generalized jackknife—finite samples and subsample sizes. (Submitted to J. Amer. Stat. Assoc.)
627. Shelton, D. W. and Weber, L. J. 1981. Quantification of the joint effects of mixtures of hypatotoxic agents: Evaluation of a theoretical model for mice. Environmental Research 26:33–41.
628. Shirazi, M. A. 1983. Alternative end points and calculation procedures to analysis of bio-assay data. Technical Report, presented at the Annual Aquatic Toxicity Workshop, Halifax, N.S., Nov. 7–10.
629. Shirley, E. 1977. A non-parametric equivalent of Williams' test for contrasting increasing dose levels of treatment. Biometrics 33:386–9.
630. Shorack, G. R. 1966. Graphical procedures for using distribution-free methods in the estimation of relative potency in dilution (direct) assays. Biometrics 22:610–9.
631. Shuster, J. J. 1968. On the inverse Gaussian distribution. J. Amer. Stat. Assoc. 63:1514–6.
632. Shuster, J. J. and Dietrich, F. H. 1976. Quantal response assays by inverse regression. Commun. in Stat. A, 5(4):293–305.
633. Shuster, J. J. and Yang, M. C. K. 1976. A distribution-free approach to quantal response assays. Can. J. Stat. 3(1):57–70.
634. Sibuya, M. 1962. On a model in probit analysis. Annals of the Institute of Statistical Mathematics 13:251–7.
635. Sielken, R. L. Jr. 1981 Re-examination of the ED01 study: risk assessment using time. Funda. Appl. Toxicol. 1:88–123.
636. Sim, D. A. 1984. Estimation of a time dependent LD50 using Cox's proportional hazards model. Paper for Biometrics Conference, Aug. 15, 1984, Philadelphia. (See Abstract in Biometrics 41(1):332.)
637. Singh, M. and Dey, A. 1985. A note on incomplete block designs for symmetrical parallel line assays. (Submitted to Communications in Statistics, 9p.)
638. Singhal, R. L. and Thomas, J. A. (Editors) 1980. Lead Toxicity. Baltimore: Urban and Schwarzenberg.
639. Skelley, D. L., Brown, L. and Besch, D. 1973. Radioimmunoassay. Clinical Chemistry 19:146–86.
640. Small, R. D., Burton, D. T., Capizzi, T. P., Hall, L. W. and Margrey, S. L. 1979. A method for analyzing data from multifactor toxiocological experiments with repeated measurements. Contributed Paper, Biometric Society Meeting April 9, 1979, 13p.
641. Smillie, K. W. 1974. APL/360 with Statistical Applications. Don Mills: Addison-Wesley.
642. Smith, K. C., Savin, N. E. and Robertson, J. L. 1984. A Monte Carlo comparison of maximum likelihood and minimum chi square sampling distributions in logit analysis. Biometrics 40(2):471–82.

180   Bibliography

643. Smith, S. W. 1981. The logit log transform from chous median effect principle. Ligand Quarterly 4(1):47.
644. Smith, W. W. and Cornfield, J. 1958. Extending the range of dose-effect curve for irradiated mice. Science 128:473–4.
645. Smythe, R. T., Krewski, D. and Murdoch, D. 1983. Ths use of historical control information in dose response problems in carcinogenesis. Report given at the Stat. Soc. of Can. Conference, June, 11p.
646. Snee, R. D. and Irr, J. D. 1981. Design of a statistical method for the analysis of mutagenesis at the hypoxanthine-quanine phosphoribosyl transferase locus of cultured Chinese hampster avary cells. Mutation Research 85:77–93.
647. Sokal, R. R. and Rohlf, F. J. 1969. Biometry. San Francisco: Freeman.
648. Sowden, R. R. 1971. Bias and accuracy of parameter estimates in a quantal response model. Biometrika 58:595–603.
649. Sowden, R. R. 1972. On the first-order bias of parameter estimates in quantal response model under alternative estimation procedures. Biometrika 59(3):573–9.
650. Spearman, C. 1908. The method of 'right and wrong cases' ('constant stimuli') without Gauss's formulae. Brit. J. of Psychology 2:227–42.
651. Spezzaferri, F. 1985. A note on multivariate calibration experiments. Biometrics 40(1):267–72.
652. Sprague, J. B. 1964. Lethal concentrations of copper and zinc for young Atlantic salmon. J. Fish. Res. Board Can. 21:17–26.
653. Sprague, J. B. 1969. Measurement of pollutant toxicity to fish: I. Bioassay methods for acute toxicity. Water Research 3(7):793:821.
654. Sprague, J. B. 1970. Measurement of pollutant toxicity to fish: II. Utilizing and applying bioassay results. Water Research 4:3–32.
655. Sprague, J. B. 1971. Measurement of pollutant toxicity to fish: III. Sublethal effects and "safe" concentrations. Water Research 5:245–66.
656. Sprague, J. B. 1973. The ABC's of pollutant bioassay using fish. In Biological Methods for the Assessment of Water Quality, American Society for Testing and Materials, STP 528:6–30.
657. Sprague, J. B. and Logan, W. J. 1979. Separate and joint toxicity to rainbow trout of substances used in drilling fluids for oil exploration. Envir. Pollution 19:269–81.
658. Sprague, J. B. and Ramsey, B. A. 1965. Lethal levels of mixed copper-zinc solutions for juvenile salmon. J. Fish. Res. Board Can. 22:425–32.
659. Sprague, J. B. and Vogels, J. 1976. Watch the Y in bioassay. Proceedings of the 1976 Aquatic Toxicity Workshop, p 107–18.
660. Spurr, H. W. Jr. and Main, C. E. 1979. Application of probit analysis to foliar disease biocontrol. Phytopathology 69(5):537.
661. Srivastava, M. S. and Carter, E. M. 1983. An Introduction to Applied Multivariate Statistics. New York: North-Holland.
662. Srivastava, M. S. and Khatri, C. G. 1979. Introduction to Multivariate Analysis. New York: North-Holland.
663. St. Pierre, J., Cadieux, M., Guerault, A. and Quevillan, M. 1976. Statistical tables to detect significance between frequencies in two small samples, with particular reference to biological assays. Revue Canadienne de Biologie 35(1):17–23.
664. Stanley, J. 1963. The Essence of Biometry. Montreal: McGill Un. Press.
665. Stephan, C. E. 1976. Methods for calculating an LC50. Preprint of a talk presented at A.S.T.M. Meeting, Oct., 1976. (From National Water Quality Laboratory, Duluth, Minnesota.)
666. Stevens, W. L. 1948. Control by gauging. J. Roy. Stat. Soc., B, 10:54–108.
667. Stevens, S. S. 1961. Is there a quantal threshold? In: Sensory Communication. W. Rosenblith (ed), Cambridge Mass.: MIT Press, p. 806–13.
668. Strand, A. L. 1930. Measuring the toxicity of insect fumigants. Indust. Eng. Chem., Analytical Edition 2:4–8.
669. Strenio, J. F., Weisberg, H. I. and Bryk, A. S. 1983. Empirical Bayes estimation of individual growth-curve parameters and their relationship to covariates. Biometrics 39(1):1–12.
670. Sun, Y. P. 1957. Bioassay of pesticide residues. Advances in Pest Control Research 1:449–96.
671. Swan, A. V. 1969. The reciprocal of Mill's ratio. Applied Statistics 18:115–6.
672. Swann, W. H. 1964. Report on the development of a new direct search method of optimization. Res. Note No. 63/4, I.C.I. Central Instr. Lab.
673. Tachibana, S. 1971. The bioassay of secretin in the rat. Japanese J. of Pharmacology 21(3):325–36.
674. Tallarida, R. J. and Jacob, L. S. 1979. Dose-response Relation in Pharmacology. New York: Springer-Verlag. (See ch. 4.)
675. Tallarida, R. J. and Murray, R. B. 1981. Manual of Pharmacologic Calculations with Computer Programs. New York: Springer-Verlag.
676. Tarone, R. E 1975. Tests for trend in life table analysis. Biometrika 62(3):679–82.
677. Tattersfield, F. and Gimingham, C. T. 1927. Studies on contact insecticides. Part V: The toxicity of the amines and N-heterocyclic compounds to *Aphis rumicis*. Annals of Applied Biology 14:217–39.

678. Tattersfield, F. and Potter, C. 1943. Biological methods of determining the insecticidal values of pyrethrum preparations (particularly extracts in heavy oil). Annals of Applied Biology 30:259–79.
679. Terza, J. W. 1985. Ordinal probit: a generalization. Communications in Statistics, Theory and Methods 14(1):1–12.
680. Thomas, D. C. 1983. Nonparametric estimation and tests of fit for dose-response relations. Biometrics 39(1):263–8.
681. Thomas, D. R. and Grunkemeier, G. L. 1975. Confidence interval estimation of survival probabilities for censored data. J. Amer. Stat. Assoc. 70(352):365–71.
682. Thomas, H. A. and Bradley, E. L. 1975. Feeding deterrents for the *Pales* weevil in a laboratory bioassay. J. of Econ. Entom. 68(2):147–9.
683. Thompson, C. H., Bratcher, T. L. and Kodell, R. L. 1978. The analysis of dose-response data with serial sacrifices. Paper presented at the Joint Statistical Conference, San Diego, Aug. 14–18, 1978.
684. Thompson, G. H. 1914. The accuracy of the $\phi(\gamma)$ process. British J. of Psychology 7:44–55.
685. Thompson, G. H. 1919. A direct deduction of the constant process used in the method of right and wrong cases. Psychol. Rev. 26:454–64.
686. Thompson, W. R. 1947. Use of moving averages and interpolation to estimate median-effective dose. Bacteriological Review 11:115–45.
687. Thompson, W. R. and Weil, C. S. 1952. On the construction of tables for moving-average interpolation. Biometrics 8:51–4.
688. Thoni, H. 1967. Transformation of variables used in the analysis of experimental and observational data. Tech. Rep. No. 7, Statistical Laboratory, Iowa State Univ., Ames, Iowa.
689. Thorburn, D. 1967. Some asymptotic properties of jackknife statistics. Biometrika 63(2):305–13.
690. Thurston, R. V., Russo, R. C. and Smith, C. E. 1978. Acute toxicity of ammonia and nitrate to cutthroat trout fry. Trans. Amer. Fish. Soc. 107:361–8.
691. Tiede, J. J. 1976. Robust calibration and radioimmunoassay. Ph.D. Thesis, SUNY at Buffalo.
692. Tiede, J. J. and Pagano, M. 1977. The application of robust calibration to radioimmunoassay. Tech. Rep. No. 52, Div. of Stat. Sc., SUNY at Buffalo.
693. Ting, C. C. 1977. Comparison of the mass action law with the power law, the probit law and the logit law in dose effect analyses. Pharmacologist 19(2):165.
694. Tocher, K. D. 1949. A note on the analysis of grouped probit data. Biometrika 36:9–17.
695. Tomarelli, P. M. and Bernhart, F. W. 1962. Biological assay of milk and whey protein compositions for infant feeding. J. of Nutrition 78(1):44–50.
696. Trevan, J. W. 1927. The error of determination of toxicity. Proc. Roy. Stat. Soc., B, 101:483–514.
697. Trivedi, R. C. and Dubey, P. S. 1978. Evaluation of toxicity of some industrial wastes to fish by bioassay. Envir. Pollution 17(1):75–80.
698. Tsay, J., Chen, I. and Heminger, L. 1978. Application of censored log-normal model to a radioimmunoassay of thyroid stimulating hormone. Tech. Rep., Univ. of Cincinnati Medical College, Cinc.
699. Tsuji, K., Elfring, G. L., Crain, H. H. and Cole, R. J. 1967. Dose response curve linearization and computer potency calculation of turbidimetric microbiological vitamin assays. Applied Microbiology 15:363–7.
700. Tsutakawa, R. K. 1967. Random walk design in bio-assay. J. Amer. Stat. Assoc. 62:842–56.
701. Tsutakawa, R. K. 1967. Asymptotic properties of the block up-and-down method in bio-assay. Annals of Mathematical Statistics 62:1822–8.
702. Tsutakawa, R. K. 1972. Design of experiment for bioassay. J. Amer. Stat. Assoc. 67:584–90.
703. Tsutakawa, R. K. 1980. Selection of dose levels for estimating a percentage point of a logistic quantal response curve. Applied Stat. 29(1):25–33.
704. Tsutakawa, R. K. 1982. Statistical methods in bioassay. In Encyclopedia of Statistical Sciences, Vol. 1 (Eds. S. Kotz, N.L. Johnson and C. B. Read) p. 236–43. N.Y: Wiley.
705. Tukey, J. W. 1958. Bias and confidence in not-quite large samples (abstract) Annals of Mathematical Statistics 29:614.
706. Turner, N. 1955. Tests for type of action of hydrocarbon insecticides applied jointly. Connecticut Agr. Exp. Station Bulletin, No. 594.
707. Van Eeden, C. 1960. On distribution-free bio-assay. Proceedings of the Symposium on Quantitative Methods in Pharmacology, Leiden, 206–10.
708. Van Ryzin, J. and Rai, K. 1980. The use of quantal data to make predictions. In: The Scientific Basis of Toxicity Assessment. Ed.: H. R. Witschi. New York: Elsevier.
709. Van Strik, R. 1960. A method of estimating relative potency and its precision in the case of semi-quantitative response. Symposium on Quantitative Methods in Pharmacology, Leiden.
710. Vandenberg, J. S. and Soper, R. S. 1979. A bioassay technique for *Entomophthora sphaerosperma* on the spruce budworm, *Choristoneura fumiferana*. J. Invert. Pathology 33:148–54.
711. Venter, J. H. 1963. On stochastic approximation methods. Ph.D. Thesis, Univ. of Chicago.
712. Vigers, G. A. and Maynard, A. W. 1977. The residual oxygen bioassay: a rapid procedure to predict effluent toxicity to rainbow trout. Water Research 11(4):343–6.

713. Vincent, L. E. and Lindgren, D. L. 1975. Toxicity of phosphine and methyl bromide at various temperatures and exposure periods to the four metamorphic stages of *Trogoderma variabile*. J. of Economic Entomology 68(1):53–6.

714. Viveros, R. and Sprott, D. A. 1985. Maximum likelihood quantal response bioassay. To appear in Symposia in Statistics, Univ. of Western Ont.

715. Vivian, S. R. and LaBella, F. S. 1971. Classical bioasssay statistical procedures applied to radioimmunoassay of bovine thyrotropin growth hormone and prolactin. J. of Clinical Endrocinology and Metabolism 33(2):225–33.

716. Vølund, A. 1978. Application of the four-parameter logistic model to bioassay: comparison with slope ratio and parallel line models. Biometrics 34:357–65.

717. Vølund, A. 1980. Multivariate bioassay. Biometrics 36:225–36.

718. Vølund, A. 1980. Dose-response model and their applications. Draft Notes, 45p.

719. Vølund, A. 1982. Combination of multivariate bioassay results. Biometrics 38:180–90.

720. Wadley, F. M. 1967. Experimental Statistics in Entomology. Graduate School Press, U.S. Dept. of Agriculture, Washington, D.C.

721. Wahrendorf, J. and Brown, C. C. 1980. Bootstrapping a basic inequality in the analysis of joint action of two drugs. Biometrics 36:653–657.

722. Walker, P. J. 1966. A method of measuring the sensitivity of trypanosomes to acriflavine and trivalent tryparsamide. J. of Genetics and Microbiology 43:45–58.

723. Walsh, N. N. 1985. Analysis of Parabolic Bioassays. M.Sc. Thesis, Dept. of Math. and Stat., Un. Of Guelph. 126p.

724. Walsh, M. N., Hubert, J. J. and Carter, E. M. 1985. Estimation methods in parabolic bioassays. To appear in the Proceedings of Symposia in Statistics, Unn. of Western Ont.

725. Ware, J. H. and Louis, T. A. 1983. Statistical problems in environmental research. Can. J. Stat. 11(1):51–70.

726. Wedemeyer, G. A. and Nelson, N. C. 1975. Statistical methods for estimating normal blood chemistry ranges and variances in rainbow trout (*Salmo gairdneri*), Shasta strain. J. Fish. Res. Board Can. 32(4):551–4.

727. Weil, C. S. 1952. Tables for convenient calculation of median effective dose (LD50 or ED50) and instructions for their use. Biometrics 8:249–63.

728. Weil, C. S. 1972. Statistics vs. safety factors and scientific judgement in the evaluation of safety for man. Toxicol. Appl. Pharmacol. 21:454–63.

729. Weil, C. S., Carpenter, C. F. and Smyth, H. F. Jr. 1953. Specifications for calculating the median effective dose. Amer. Indust. Hygiene Assoc. J. 14:200–6.

730. Wesley, M. N. 1976. Bioassay: estimating the mean of the tolerance distribution. Tech. Rep. No. 17, Div. of Biostat., Stanford Univ.

731. Wetherill, G. B. 1963. Sequential estimation of quantal response curves (with discussion). J. Roy. Stat. Soc., B, 25:1–48.

732. Wetherill, G. B. 1966. Sequential Methods in Statistics. London: Methuen. (Second Edition, 1975, London: Chapman and Hill.)

733. Wetherill, G. B., Chen, H. and Vasudeva, R. B. 1966. Sequential estimation of quantal response curves: a new method of estimation. Biometrika 53:439–54.

734. White, R. F. and Graca, J. G. 1958. Multinomially grouped response times for the quantal response bioassay. Biometrics 14:462–88.

735. Whitlock, J. H. and Bliss, C. I. 1943. A bioassay technique for antihelmintics. J. of Parasitology 29:48–58.

736. Whittle, P. 1957. Curve and periodogram smoothing. J. Roy. Stat. Soc., B, 19:38–47.

737. Whittle, P. 1958. On the smoothing of probability density functions. J. Roy. Stat. Soc., B, 20:334–43.

738. Williams, D. A. 1971. A test for differences between treatment means when several dose levels are compared with a zero dose control. Biometrics 27:103–17.

739. Williams, D. A. 1972. The comparison of several dose levels with a zero dose control. Biometrics 28(2):519–31.

740. Williams, D. A. 1973. The estimation of relative potency from two parabolas in symmetric bioassays. Biometrics 29(4):695–700.

741. Williams, D. A. 1975. The analysis of binary responses from toxicological experiments involving reproduction and teratogenicity. Biometrics 31:949–52.

742. Williams, D. A. 1978. An exact confidence region for a relative potency estimated from combined bioassays. Biometrics 34(4):659–61.

743. Williams, D. A. 1982. Extra-binomial variation in logistic linear models. Applied Statistics 31:144–8.

744. Williams, E. J. 1969. A note on regression methods in calibration. Technometrics 11:189–92.

745. Wilson, E. B. and Worcester, J. 1943. The determination of LD50 and its sampling error in bio-assay. Proc. Nat. Acad. Sciences 29:79–85.

746. Win, K. and Dey, A. 1980. Incomplete block designs for parallel-line assays. Biometrics 36:487–92.

747. Winder, C. V. 1947. Misuse of 'deduced ratios' in the estimation of median effective doses. Nature 159:883.

748. Wong, S. L. and Beaver, J. L. 1980. Algal bioassays to determine toxicity of metal mixtures. Hydrobiologia 74(3):199–208.

749. Worcester, J. and Wilson, E. B. 1943. A table determining LD50 or 50% end-point. Proc. Nat. Acad. Sci., Wash. 29:207–12.
750. Wu, C. F. J. 1985. Efficient sequential designs with binary data. J. Amer. Stat. Soc. 80(392):974–84.
751. Xu, D–Z. 1981. Parabola bioassays in the comparison of drug effects. Acta Pharmacology 2(2):73–8.
752. Yalow, R. S. and Berson, S. A. 1960. Immunoassay of endogenous plasma insulin in man. J. Clinical Invest. 39:1157–75.
753. Yalow, R. S. and Berson, S. A. 1968. General principles of radioimmunoassay. In: Radioisotopes in Medicine: In Vitro Studies. (Hages, R. L., Goswitz, F. A. and Murphy, B. E. P., eds.) U.S. Energy Commission Technical Information Centre, Oakridge, p.7–42.
754. Yates, F. 1937. The Design and Analysis of Factorial Experiments. Harpenden, England: Imperial Bureau of Soil Science.
755. Young, D. M. and Roman, R. G. 1948. Assays of insulin with one blood sample per rabbit per test day. Biometrics 4:122–31.
756. Young, W. R. 1983. Statistical estimation of relative potencies from parallel line bioassays. Tech Rep., Cyanamid Corp., 25p.
757. Young, W. R. 1983. Operating instructions for interactive joint probit program. Tech. Rep., Cyanamid Corp., 32p.
758. Zaba, B. 1979. The four parameter logit life table system. Population Studies 33(1):79–100.
759. Zar, J. H. 1974. Biostatistical Analysis. Englewood Cliffs: Prentice-Hall.
760. Zeger, S. and Brookmeyer, R. 1984. Analysis of serial measurements subject to censoring with application of bioassay. Paper for ASA Conference, Aug. 16, 1984, Philadelphia.
761. Zelen, M. 1970. Theory of Biometry (lectures). Dept. of Stat., SUNY at Buffalo.
762. Zerbe, G. O. 1978. On Fieller's theorem and the general linear model. American Statistician 32(3):103–5.
763. Zerbe, G. O. 1979. Randomization analysis of the completely randomized design extended to growth and response curves. J. Amer. Stat. Assoc. 74:215–21.
764. Zerbe, G. O. 1979. Reply (Letter). American Statistician 33:162.
765. Zerbe, G. O., Laska, E., Meisner, M. and Kushner, H. B. 1982. On multivariate confidence limits for ratios. Commun. in Stat., Theoretical Methods 11(21):2401–25.
766. Zitko, V. 1979. An equation of lethality curves in tests with aquatic fauna. Chemosphere 2:47–51.

Index